# Scientific Pioneers

## Women Succeeding in Science

Joyce Tang

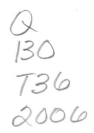

UNIVERSITY PRESS OF AMERICA,® INC.
Lanham • Boulder • New York • Toronto • Oxford

**Copyright © 2006 by**
**University Press of America,® Inc.**
4501 Forbes Boulevard
Suite 200
Lanham, Maryland 20706
UPA Acquisitions Department (301) 459-3366

PO Box 317
Oxford
OX2 9RU, UK

Library of Congress Control Number: 2005935100
ISBN 0-7618-3350-1 (paperback : alk. ppr.)

∞™ The paper used in this publication meets the minimum
requirements of American National Standard for Information
Sciences—Permanence of Paper for Printed Library Materials,
ANSI Z39.48—1992

In memory of my mother
and for my father

# Contents

# List of Tables

# Preface

Female scientific pioneers are the central focus of this book. Why should we be so interested in studying the career success of women in science? The theme of this book is an important one—some women succeeded in science in the last century, more women could succeed in this century. Knowledge of successful female scientists would provide suggestions for improving the training of women now and in the future. Stability and change is another theme of this book. This anatomy of female scientific pioneers will make contribution to both public policy debates and research on inequality. This book is timely in light of the recent remarks by Harvard's president about women's talents in science and mathematics. While many social scientists focus on the barriers inherent in the society and in the scientific profession, I seek to understand why some women have succeeded in this male endeavor. I also investigate possible reasons for the consistent finding of their remarkable successes in spite of repeated failures. I propose an explanation for this, focusing on science career as an outcome of individual attributes, structural opportunities, and institutional support. *Scientific Pioneers* provides support for the view that career advancement is a continuous, dynamic process of choice, design, and adaptation. Rather than being a passive agent, successful female scientists in this book are actively making conscious decisions on career progress in light of their own interest and circumstances. Equally important, to make sense of the strategies and tactics that these exceptional women have used, we need to bring their personal and professional environment into analysis.

*Scientific Pioneers* is a reversal of tradition in research on women and science and on scientific pioneers. Books on women in science focus on the barriers that women face in education and employment. As a result, we know more about why women continue to lag behind men in scientific achievements. The question as to why women have managed to stay in and get ahead in science despite well-documented educational and occupational barriers remains unanswered. The message of this book is the subtitle. It deals with women who succeed in science—doing what many said could not be done. But rather than concentrating on the obstacles and hardships that they face in life and in work that preoccupied other scholars, I explore the determining factors and circumstances behind their ascent in the scientific establishment.

*Scientific Pioneers* critically examines the career development of female scientific pioneers. I put together chapters in a book that includes history, theories, and analysis—derived from (a) the history of science and the sociology of science literature, (b) the works in social stratification and mobility, and (c) detailed analysis of life and career histories of ten notable figures of recent history of science. I integrate existing biographical and ethnographical data with the general themes of the book. Incorporating quotations from published accounts of these successful female scientists and significant others into discussions give flesh and meaning to the analysis. These materials, I hope, would deepen the reader's understanding of what the analysis means for careers of talented women in reality. This approach will contribute to a more comprehensive understanding of the complexity of career progress among scientific pioneers.

*Scientific Pioneers* will add to the growing literature on women in science, the sociology of science, and the dynamics of discrimination. We can use the case of female scientific pioneers to marshal support for the claim that throughout their career they have successfully translated their skills and training into excellence and success (e.g., receipt of Nobel Prize and other honors for scientific excellence). There is also evidence for the assertion that social movements in the last century have pried open the door of opportunities for educated women. But these changes have facilitated only a tiny segment of the female population to reach the top of the scientific establishment.

Chapter 1 gives a brief account of scientific pioneers, with a focus on the characteristics of highly successful scientists. This chapter sets the investigation of the making of female scientific pioneers in proper context. It begins with an overview of the indicators of career success in science. To shed light on the issues of women's invisibility among scientific elites, I also examine (a) various approaches that researchers use to differentiate between pioneers and non-pioneers, and (b) the reasons why so little attention have been given to women who have become scientific pioneers. It is followed by a discussion of the rationale as well as the background for initiating the first sociological study of female scientific pioneers. The last part of this chapter is devoted to examining the special traits and circumstances of individuals who have had outstanding scientific careers.

Chapter 2 reviews four major arguments for the science gender gap. It offers a few illustrations of the ways in which different approaches or explanations shed lights on the process of choosing science as a career choice. It sets the stage for analyzing the origins, development, and advancement of female scientific pioneers in forthcoming chapters. Drawing from the literature on two disciplines—the history of science and the sociology of science, this chapter gives a more comprehensive view of women's relative standing in the scientific establishment. The biological approach underscores the influence of ascribed characteristics on educational and career outcomes. The individual choice approach revolves around personal interest in academic and career choices. Both structural and institutional approaches attribute career success or failure to external forces, which could be opportunities as well as constraints. However, the institutional approach focuses on the differential impact of family, schools, and profession on career advancement. Besides comparing their strengths and limitations, when appropriate, I highlight major differences, or draw parallels, between certain approaches. I also discuss how each of these approaches predicts the career achievements of female scientists. Throughout Chapter 2, I lay out relevant issues to be addressed in subsequent chapters.

Not everyone who engages in scientific endeavor will ultimately be recognized as scientific pioneers. The question central to Chapter 3 is: How is it possible to succeed in science in the twentieth century? The purpose of this chapter is to highlight the similarities and differences in the background and characteristics of ten female scientific pioneers. The theme is: male and female scientific pioneers may have similar traits underlying their success, although they have to overcome numerous (yet different) barriers at different stages of their careers.

In addition to having a passion for science, female scientific pioneers share many attributes that are considered as absolutely essential in doing pioneering work. They are high achievers who have succeeded in overcoming numerous obstacles in family, school, and work. Very early in life and career, their actions and behavior reflect intelligence, optimism, courage, hard work, discipline, independence, enthusiasm, drive, competitiveness, ability to focus and concentrate, ambition, dedication, confidence, assertiveness, creative thinking and work styles, risk-taking, and perseverance. Many of them have shown exceptional academic abilities in childhood. Their drive and determination are consistently demonstrated in their choice of subjects for study, selection of topics for investigation, as well as their academic and career plans. With or without any external support, many of them have been able to deal with frustrations and disappointment by teachers, peers, colleagues, or employers. Most importantly, they tend to perceive obstacles merely as humps to get over with rather than stumbling blocks in order to get to the next stage of work or career.

Chapter 4 explores the conditions that make the production of female scientific pioneers possible. Many women who possess impressive attributes to do pioneering work as described in preceding chapter do not become scientific pioneers. So why are men and women with similar traits and qualities differentially rewarded for their contributions to science? Other factors can shape one's aspirations for science career and scientific development. Being outstanding in ability, effort, and style alone is not sufficient to propel a talented woman to the forefront of science. It takes a change of circumstances. A set of structural forces must be in place to open up new possibilities for women. This means that it becomes necessary for the society to allow for women's greater participation in science. Throughout the chapter, the discussions underscore the fact that the economic and political upheavals in the mid-nineteenth and mid-twentieth centuries had provided newcomers (such as women) with the opportunities to pursue pioneering work in science. I classify the structural shifts that precipitate women's entry to the scientific establishment into three categories: cultural, economic, and political. The discussions illustrate the complexity of the interaction between these collective transformations.

Chapter 5 examines three dimensions of institutional influences on career advancement: family, schools/institutes, and science. These forces combine to elicit career success in science. Family is the institution that has the most enduring impact on career aspirations and career development of female scientific pioneers. Support from academic or research institutions is critical in their career development—as a training ground, a major source of research support, and a gateway for networks. The rise of professionalism in science in the late nineteenth and twentieth centuries presented both challenges and opportunities to female scientists. Aside from individual talents and effort, discussions in this chapter emphasize the importance of being brought up and trained in a stimulating environment (family and school/institutes) for the advancement of female scientists. Drawing from their life and career histories, I demonstrate that talented, hardworking women were able to develop new ideas and make significant discoveries, because they happened to be in places that were intellectually stimulating and resourceful as well as at a time when it was ripe for scientific breakthroughs.

Chapter 6 revisits the central ideas and main arguments in the field of sociology of science. The question central to this chapter is: What do the experiences of female scientific pioneers tell us about the normative structure of science? Not all norms in science are gender-blind. Women in particular have difficulty in adhering to these norms. As a result of trial and error, female scientific pioneers have successfully navigated through the male culture of science. Based on the analyses and discussions in preceding chapters, I argue for the existence of norms and counternorms in the scientific establishment. Collectively, these (counter)norms impact positive and negatively on the career development and advancement of female scientists. I also highlight the adaptive strategies used by female scientific pioneers to advance their careers.

The concluding chapter gives a summary of analyses presented in the book by addressing several key questions: Are female scientific pioneers born or made? What motivates them? How do their backgrounds and family relationships shape their career aspirations and career trajectories? What brings them to the ranks of scientific pioneers? How do they gain recognition? Who supports these women in becoming scientific pioneers? Why did the scientific establishment and society for the first time increasingly let women join the ranks of scientific elites? Answers to these questions demystify the production of female scientific pioneers. Aside from synthesizing the findings and assessing the implications for policy debates, theoretical development, and research, I speculate on the prospects of women in science: What does the future hold for women in science? What can be done to improve their current status in science. The discussions lay the groundwork for comparative studies in other male endeavors.

This book is the result of numerous conversations with students, colleagues, and peers at Queens College and elsewhere about the achievements of women in science over the years. I have also benefited immensely from the thoughtful comments and suggestions from anonymous reviewers. I am indebted to Donny Tang and Earl Smith for their support and perceptive comments. Arlene Robinson offered professional editing of the manuscript with care and enthusiasm. This research was supported in part by funding from the Professional Staff Congress—City University of New York (PSC-CUNY) Research Foundation under grants #61648, #62729, #63693, #64684, and #65724. However, the findings and conclusions do not necessarily represent the official opinion of the funding agency.

*Chapter One*

# Scientific Pioneers

People choose science for a variety of reasons. Chief among them are *puzzlement* (curiosity), *the gold quest* (financial rewards), and *the blue ribbon* (recognition). However, few have made it to the top of the scientific establishment and achieved these goals.

So, what does it take to reach the pinnacle of the field? Scientific pioneers are often considered the first among equals. What sets pioneers apart from nonpioneers? In particular, what does it take for a woman to become a scientific pioneer? Even with extraordinary abilities, it is difficult enough for a man to succeed in the highly competitive fields of science. Women have encountered all the hurdles men have faced over the years, and more. Even with these obstacles, a handful of women have succeeded. How did these women make such accomplishments? Answers to these questions would be extremely valuable to young women aspiring to scientific careers.

## IN SEARCH OF SCIENTIFIC PIONEERS

While it might be difficult to become a pioneer, it is relatively easy to recognize one. Their discoveries or innovations are not merely important, but recognized as such; their breakthroughs actually change the direction of research in their fields. Scholars have used terms such as *precursors*, *geniuses*, *creators*, *stars*, and *elites* to label individuals with outstanding careers or exceptional achievements in the field of science (Howe 1999; Michalko 1998; Ochse 1990; Simonton 1999; Stephan and Levin 1993; Zuckerman 1996).

Sociopsychological studies of those showing scientific excellence have made more precise distinctions based on career orientations, scientific reputations, developmental stages, and scientific approaches. Joseph Hermanowicz, who examined social order in the science professions, stated that academic scientists could be found in one of three spheres:

*1. Elites* are most committed to as well as most successful at research.
*2. Communitarians* are most devoted to teaching.
*3. Pluralists* are similar to elites or communitarians, or something in between.

Results of his study showed that there was more similarity in career orientations between pluralists and elites than between pluralists and communitarians (Hermanowicz 1998:201–204).

Perceived quality and visibility of research contributions have a bearing on a person's reputation within the scientific community. To differentiate between scientists, Jonathan Cole developed a typology of scientific reputation that categorized scientists into four types:

*1. Prominent:* Their work is widely known and highly regarded in the field.
*2. Notorious:* Their work is widely known but not highly regarded.
*3. Esteemed*: Their work is known as important among peers who know of it, but they have either chosen not to seek fame or have not yet earned it.
*4. Invisible:* Their work is neither widely known nor highly regarded.

In spite of his characterizations, Cole reminds us that reputational standing should be viewed as a continuum: Due to the competitive nature of science, movement from one type of reputation to another is common in the scientific community (Cole 1979:102–106).

Scientists can also be differentiated by their approach to conducting scientific investigations. Dean Keith Simonton identified three different ways scientists study nature:

*1. Pioneers* or *precursors* venture into unknown or unexplored domains.
*2. Advancers* are the majority of scientists, who try to build on an existing body of knowledge. Advancers can be distinguished by their degree of contribution to the prevailing paradigm:
   a. *Extenders* "add more pieces to a single configuration in the field."
   b. *Synthesizers* "manage to bridge two (or more) hitherto separate configurations of letters."
*3. Revolutionaries* rearrange the pieces to generate more explanatory power over previous configurations (Simonton 1988:181–183).

Surprisingly, Simonton observes that pioneers seldom earn as much credit for their work as advancers do. The reason is that pioneers tend to pursue independent work in undefined or unknown domains; their discoveries or ideas tend to be incomplete. As a result, their initial ideas are less likely to be fully accepted into the prevailing paradigm (Simonton 1988:125–126).

The distinctions among scientists proposed by Hermanowicz, Cole, and Simonton focused on various aspects of their work and their impact on the scientific world. Michael Howe used a life-course approach to categorize scientists, and pinpointed several factors that affect the development of scientific talents at different stages of life. Howe found that a large number of "creative geniuses" grew up in a stimulating home environment, were prodigies in midchildhood, and demonstrated the capacity of creative achievement in adulthood. Nobel laureates Albert Einstein and Marie Curie, musicians Wolfgang Amadeus Mozart and Yehudi Menuhin, thinker John Stuart Mill, and mathematician Norbert Wiener fit into this category. This framework underscores the interaction of a combination of sociopsychological factors in the development of scientific talents, and suggests that the making of scientific pioneers is most likely under certain circumstances. Through his study, Howe reminds us to pay attention to key determinants and events and how each works during the process of developing scientific talents.

The preceding discussions suggest there can be multiple distinctions of scientists in different contexts. However, neither Hermanowicz's categorization of scientists nor Cole's typology of scientific reputation can help us clearly identify scientific "pioneers." It is conceivable that individuals who are devoted to research (*elites*), or teaching (*communitarians*), or to both (*pluralists*) can make significant discoveries or innovations. Similarly, scientific pioneers can be found among "prominent," "notorious," "esteemed," or "invisible" scientists under different circumstances. Conversely, one can be a pioneer in science without gaining eminence or having a successful career. This is evident in the historical underrecognition or underreward of scientific discoveries made by women and minorities. It is imperative to note that terms such as "eminent," "pioneer," and "successful" are related but *not* interchangeable, even though these qualities tend to overlap with one another. Also, doing pioneering work does not guarantee that one will gain prominence or have an outstanding career. For most scientists, success does not necessarily mean receiving the Nobel Prize. Whether a scientist is doing good work, helping others, or making the world a better place to live should also be taken into consideration.

Simonton offers a simple, yet valuable definition of a scientific pioneer based on different modes of scientific investigation. Howe's developmental approach identifies critical factors in the development of scientific talents in

early childhood, adolescence, and adulthood. (I will use Simonton's and
Howe's approaches to guide discussion in this and forthcoming chapters.) Yet
in spite of these identifying factors, the scientific establishment remains a patriarchal environment, and men constitute an overwhelming majority of those
considered pioneers in science. Women have rarely been considered (or perceived) by their peers and others as scientific pioneers. There is ample evidence in recently published profiles of scientists to substantiate this claim of
women's invisibility among scientific elites. Chief among them are:

*The popular perception of science as a male-dominated field.* Science is
generally considered a male-dominated field among men and women alike.
When we asked ordinary people, including high school and college students,[1]
to list a few famous scientists, their responses usually included inventors or
engineers considered household names (for example, Thomas Edison, Albert
Einstein, and Isaac Newton) and renowned figures in the scientific community (such as Enrico Fermi, Richard Feynman, and James Watson). Only oc-
casionally did we come across the names of a few women on their lists. The
perception that women are not leading contributors to science was supported
by the results of a study of male and female scientists' reputations. Women
were infrequently regarded as influential by their male and female peers in
three fields: sociology, psychology, and biology. Women in sociology and
psychology, however, were more likely than their male counterparts to name
at least one woman as a leading contributor. No gender differences in naming
pattern were found among biologists (Cole 1979:102–103). These findings
suggest that women's work is not highly known or regarded in the scientific
community, and constitutes a major challenge for talented female scientists
who want to join the ranks of scientific elites.

*The perpetuation of science (or creative activities) in media and academe
as masculine pursuits.* Images and stereotypes of scientific or creative activities as a male endeavor prevail in advertising, television, and movies (LaFollette 1988). Rachel Carson was the only woman among 20 "greatest scientists
of the twentieth century" profiled in the *TIME 100* issue (1999). Five of the
72 individuals included in Howe's personalia were women—Anne Brontë,
Charlotte Brontë, Emily Brontë, Marie Curie, and Mary Anne Evans (also
known as George Eliot) (1999:206–211). Three women—Marie Curie,
Gertrude Elion, and Lynn Margolis—were on Simmons' (1996) list of the
"100 most influential scientists." Due to differences in the criteria used in
compiling the list of notable scientists, the lists of names may vary. But there
is one thing in common among those profiled: They were and still are predominantly Caucasian males of European descent. It is imperative to note that
while this is true of many lists of famous scientists, this is probably not true
if one examines lists of the most highly cited individuals. Nonetheless, stud-

ies of great scientists rarely considered women for comparison and discussion. With the passage of time, there may be more progressive changes in women's status in science. This is why I argue that the first two reasons and the next one are related. One cannot talk about stereotypes and perceptions of science and scientists without considering the recency of women's entry to science.

*Women have been allowed to participate in science only in recent times.* Since science in the United States and elsewhere has been and continues to be a male domain, women have made inroads only in recent decades. According to the norm of universalism, credit and recognition should be attributed to individuals based on merits (Merton 1973), yet historical restrictions to women's participation in science run counter to the norm of universalism. Women's notable absence among scientific elites is another working example. The Nobel Prize is widely considered the equivalent of baseball's or football's Hall of Fame and Hollywood "Stars," and women constitute 2.39% of the Nobel Prize winners in three scientific fields. The principle of universalism has been directly challenged by others as well. In his analysis of the Nobel Prizes in science, Moulin was highly skeptical of the objectivity of its awarding mechanism:

> It cannot be denied that factors extraneous to the purely scientific value of the candidate's achievement may influence the making of the award. . . . [For example,] there may be rivalries of personalities and cliques. . . . [As a result,] the Nobel Prize-winners do not include all the eminent men of their country and century, or even comprise the most eminent (1955:247–248).

This statement raises questions about the reward structure of science as well as its implications for mobility in science, for women in particular (e.g., How do men and women with comparable talents fare in competing for limited resources and recognition? How likely would they be competing with one another on a level playing field?). Moulin's assertion underscores a disjuncture between the theoretical and empirical formulations of universalism.

The most straightforward evidence for Moulin's claim comes from an observation by a 1948 Nobel laureate in chemistry as well as a former chair of the Nobel Foundation. This is how Arne Wilhelm Kaurin Tiselius (1902–1971) characterized the gap between the principle and practice of picking winners:

> You cannot in practice apply the principle that the Nobel Prize should be given to the person who is best; you cannot define who is best. Therefore, you are left with the only alternative: to try to find a particular worthy candidate (Bernhard 1997:40).

This admission suggests that objective judging is hard to come by even with good intentions. The outcome at best reflects a subjective application of universal standards. This also means that we cannot explain women's lower achievements in science without understanding the role of the standard-bearers: namely, gatekeepers of science (e.g., What role do gatekeepers play in the production of female scientific pioneers?). This and related issues will be addressed in forthcoming chapters.

## MOBILITY IN SCIENCE: THE CASE OF FEMALE SCIENTIFIC PIONEERS

We have gained a better understanding of stratification in science from scientific pioneers; however, this book is a sociological study of some extraordinary women in a variety of scientific fields. We can learn *much more* about mobility in science from female scientific pioneers. For reasons that have not yet been studied thoroughly, few women have made it to the top, and many factors (e.g., prejudice, lack of access to resources) might be responsible for their relative position in science. This brings us to an important question: How do female scientists circumvent enormous barriers and turn disadvantages into advantages? To this end, we will analyze a subgroup of scientific pioneers, focusing on female scientists in the United States and Europe (to control for cultural and geographical differences). These scientists' lives and career histories constitute a rich but largely unanalyzed data source.

This microlevel study of female scientific pioneers, where we narrow our scope to a small number of world-renowned female scientists, also has methodological implications for research on women in science. Conducting a case study of female scientific pioneers allows us to learn about the fine details of career development and advancement in science, an advantage over the bulk of sociological research on scientists, which are at the level of macrolevel analyses (Allison 1980; Cole 1979; Crane 1965; Long 1992; Preston 1994; Reskin 1979; Wright 1997). In addition, quantitative data is more useful than qualitative data in showing where women stand in relation to men in the scientific profession (e.g., gaps in earnings and promotion rates). This allows researchers to analyze tangible qualifications such as education, work experience, and specialized training. Based on this information, we can make sense of a person's career growth. However, quantitative information does not always adequately capture the full range of experiences of women in science, since many other attributes are intangible and nonlinear, and therefore hard to infer from a membership directory, survey, or interview. But often, these special traits define the term "scientific pioneers." Society can train a talented man or woman in the principles and practices of science. It is very difficult, however, to train someone to have the drive, discipline, and energy to continuously test new ideas. The large number of factors shaping the course of career development underscores a serious methodological problem in empirical studies on female scientists. As a result, quantitative studies seldom provide a substantive understanding of research on female scientists.

I draw from the literature in two disciplines: the history of science as well as the sociology of science. The former discipline was developed in the 1920s and 1930s, while the latter emerged in the 1970s. Scholars in both fields have examined the status and progress of women in science by concentrating on the history of women in science and the sociology of women in science (e.g., Abir-Am and Outram 1987; Ainley 1990; Cole 1979; Reskin 1976; Rossiter 1982, 1995; Zuckerman, Cole, and Bruer 1991). Together, they provide a rich source of information on the changing role of women in science from a sociohistorical perspective.

This book examines the lives and work of ten women. Who are these people? As early pioneers of science or leading contributors to science, Marie Curie, Irene Joliot-Curie, Margaret Mead, Barbara McClintock, Maria Goeppert-Mayer, Rachel Carson, Rita Levi-Montalcini, Dorothy Hodgkin, Rosalyn Yalow, and Fay Ajzenberg-Selove all left an indelible mark on the scientific community. Since, as noted earlier, success does not simply mean winning the Nobel Prize, I included notable women who received other scientific honors or recognition for their contribution to the scientific profession. As shown in Table 1, the sample includes seven Nobel laureates and three scientists from various fields. They also worked in different settings that were completely dominated by men and the male culture at the time: university, research institute, hospital, research lab, museum, and government. How did I find them? These female scientists were selected for this study based on the following criteria:

1. Their ideas or discoveries, in one way or another, transformed the direction of a particular line of research field, or their work changed different aspects of the social, natural, or physical sciences.
2. They received substantial recognition for their contribution to society or to the scientific community.
3. They emerged as role models for young scientists, and/or, drawing from (auto)biographies or ethnographic studies, their hindsight and insights became a useful source of information for in-depth analysis.

This wide-ranging set of definitions allows us to include social scientist Margaret Mead, considered by many as an influential figure in the field of cultural anthropology. Marine biologist and science writer Rachel Carson

**Table 1. Sample of Female Scientific Pioneers**

| *Name* | *Fields* | | *Recipient of Nobel Prize/Honor(s) (Award Year) *Co-recipient* |
|---|---|---|---|
| **Marie Curie** | (1867–1934) | Physicist & Radiochemist | Physics (1903)*; Chemistry (1911) |
| **Irene Joliot-Curie** | (1897–1956) | Radiochemist | Chemistry (1935)* |
| **Margaret Mead** | (1901–1978) | Cultural Anthropologist | President of the American Anthropological Association (1960), President of the American Association for the Advancement of Science (1975) |
| **Barbara McClintock** | (1902–1992) | Geneticist | Physiology & Medicine (1983) |
| **Maria Goeppert-Mayer** | (1906–1972) | Mathematical Physicist | Physics (1963)* |
| **Rachel Carson** | (1907–1964) | Marine Biologist | National Audubon Society's medal (1963), American Geographical Society's medal (1963), Presidential Medal of Freedom (1980), Issue of Rachel Carson stamp (1981) |
| **Rita Levi-Montalcini** | (1909– ) | Neuroembryologist | Physiology & Medicine (1986)* |
| **Dorothy Hodgkin** | (1910–1994) | Crystallographer | Chemistry (1964) |
| **Rosalyn Yalow** | (1921– ) | Medical Physicist | Physiology & Medicine (1977) |
| **Fay Ajzenberg-Selove** | (1926– ) | Nuclear Physicist | American Physical Society's Nicholson Medal for Humanitarian Service (1999) |

does not seem to fit in this group as well as the others do because of her unique background. However, her writing has had profound impact on awareness of environmental issues as well as policies in the United States and other parts of the world. Fay Ajzenberg-Selove is perhaps better known for her autobiography than for her scientific research. However, her work in nuclear physics along with her pivotal role in the professional community makes Ajzenberg-Selove a pioneer in science.

The broad selection criteria are somewhat similar to the constraints encountered by those who studied creative geniuses. In his study of people capable of remarkable achievements, Howe's choices were limited by (a) the lack of availability of detailed accounts of the individual's formative years and (b) the challenge of including those with enough commonalities for meaningful comparison, such as belonging to the same era and sharing a common culture (Howe 1999:19–20). As with Howe's study, the women selected for this study lived very different lives, in very different times and places. As examples, the educational, governmental, and social structures (e.g., marriage, family, and attitudes toward extramarital sex and same-sex relationships) differ tremendously between Poland, France, England, and the United States, and between the 1860s (when Marie Curie was born) and the 1920s (when Fay Ajzenberg-Selove was born). Despite variations in their backgrounds, cultures, and structural arrangements, I will attempt to show that there are more similarities than differences in the origins, developments, and advancements of these ten women.

Admittedly, this definition of scientific pioneers has its limitations in identifying notable female scientists (Ochse 1990; Simonton 1999). First, merits do not always dictate the allocation of rewards all of the time. Regardless of gender, not all who are exceptionally talented in science have been recognized by their peers or by the scientific community. Yet historical evidence reveals that women are disproportionately underrecognized for their contributions to science. Since women are differentially rewarded for their work, it is nearly impossible to identify all female scientists who received the credit they deserved.

Second, to a large extent, the term "scientific pioneer" is often used to refer to individuals *after* their ideas or discoveries have been tested or validated. We seldom apply this label to individuals who have demonstrated the potential, but have not yet produced works that garner public or institutional recognition. As a result, this study excludes the so-called rising stars.

Third, a scientist can be a pioneer in a neglected or currently ignored field, or work in a field not yet declared important. Women in these fields may have met with the same obstacles as their counterparts in similar fields, and the

significance or importance of their work will not be determined until the next
generation. Thus, results of this study cannot be generalized to pioneers who
are outside the mainstream of science.

In short, the sample in this study is unrepresentative of the scientific
population, because these ten women were more determined and able than the
most of the female scientific labor force. However, this limitation is un-
avoidable. We can only study the careers of pioneers among those who remain and advance in science.


# CHARACTERISTICS OF PIONEERS

Multiple factors shape a young person's aspiration for a scientific career. In
this section, I attempt to identify the special traits and circumstances of individuals who have outstanding scientific careers. There is no better place to do
so than an examination of selected works on famous scientists. Regardless of
the academic disciplines or backgrounds of the investigators, they have considered a wide range of factors (Berry 1981; Clark and Rice 1982; Gray 1962;
Gustin 1973; Hargens 1978; Simon 1974; Sonnert and Holton 1995; Tang
2005; Zuckerman 1996). Despite variations in focus, a common theme
emerges out of these rankings, biographies, or studies: One can join the ranks
of scientific pioneers, geniuses, or elites if one possesses certain qualities,
was brought up in a certain family environment, and was educated and/or
worked in certain contexts. In other words, these works underscore the multiplicity of individual, structural, and institutional forces upon an aspiring scientific pioneer's potential.


# INDIVIDUAL FACTORS

## Creative Thinking Style

Being extremely creative—"thinking out of the box"—is the formula for
making scientific breakthroughs, and scientific pioneers display unconventional thinking styles when approaching an existing or new problem. They
stand out in terms of extraordinary originality as well as speed (Hermanowicz 1998:74). In ranking the most influential scientists in the past and present,
Simmons (1996) observes that what set the one hundred scientific geniuses
apart from others was "the significance of their ideas." For the most part, they
were distinguished for "discovering new things about nature" but not for

"manipulating it for other ends" (Simmons 1996:xvi). All this suggests that scientific thinkers are both originators and adventurers.

In *Cracking Creativity: The Secrets of Creative Genius*, Michael Michalko (1998) examined the thinking styles and problem-solving strategies of the who's who in the sciences, arts, and industry. Based on a content analysis of their notebooks, correspondence, conversations, and ideas, Michalko generated nine strategies shared by creative thinkers. They "know how to see," "make their thought visible," "think fluently," "make novel combinations," "connect the unconnected," "look at the other side," "look in other worlds," "find what they are not looking for," and "awaken the collaborative spirit." In short, these strategies can be captured in two themes: (a) "seeing what no one else is seeing" and (b) "thinking what no one else is thinking." Simply put, their style is creative.

Michalko's analyses provide valuable insights into how scientific discoverers, innovators, and inventors go about their business. Unfortunately, his findings and discussions provided no information for the reader to determine if (and to what extent) male and female scientists think alike or differently, or if these traits are common among both male and female geniuses. This is a major limitation of Michalko's inquiry, especially when he underscores the importance of "thinking differently," "thinking out of the box," "how men and women think differently," and "asking not why but why not" in generating innovative or novel ideas. For instance, when discussing the concept of switching gender for the first strategy—"knowing how to see"—Michalko posed a series of challenging questions which in turn highlighted the limitation of his probe (1998:40–41). If taking on a different role (such as the role of the opposite sex)—which he calls "playing the mental game of switching genders"—might change our attention, viewpoints, or perspectives, and the way we approach a situation, we could certainly benefit from similar studies of the thinking style of female geniuses (e.g., Tosi 1975).

To date, no one has conducted a systematic study to compare the thinking styles of men and women, and we do not know if there are any parallels or variations in the thinking styles among male and female scientists. The good news is that everyone has the potential to become a creative thinker, since creative thinkers are made, not born (Howe 1999; Ochse 1990; Simonton 1999). What are believed to be spontaneous creative-thinking styles, as noted by Michalko and others, are primarily the product of learning and practice. With sufficient training, patience, and effort, a society can manufacture plenty of (male and female) Einsteins. The implicit assumption in this claim is that we can rule out a lack of creative ability as an explanation for women's lower achievements in science. Other factors may be in operation, however, to hamper the development of creativity in women.

## Diversity in Problem Selection and Problem Solution

Creative thinking styles are reflected in the ways in which a person picks a topic, approaches a problem, or conducts scientific investigations. Another example to illustrate the striking similarities and inadequacies in the portrait of scientific pioneers comes from Lewis Wolpert and Alison Richards' (1988) interviews of 13 prominent scientists. Eleven of the 13 profiled in *A Passion for Science* are British scientists. All but one is male. Nobel laureate in chemistry Dorothy Hodgkin, along with three male molecular biologists, were featured under the heading of "molecules of life." Based on their conversations with these scientists and a brief description of their work, Wolpert and Richards help us make sense of the worlds (perceived or real) in which scientists function. One recurring theme is that there is no single approach to doing good science work (theoretical/experimental, pure/applied, collaboration/competition) or to making discoveries (by chance or through imagination). It can be a combination of all of these techniques and strategies. The 13 men and women they interviewed consistently reported using new and/or divergent views or methods to choose and solve problems.

Equally important, Wolpert and Richards noted that the public's understanding of science is quite different from what goes on in a scientist's laboratory. Regardless of fields or specialties, there is a great deal of variation in how scientists generate ideas and conduct research. If this is indeed the case, as shown in *A Passion for Science*, one wonders why Wolpert and Richards did not draw on the experiences of scientists from a broader spectrum of fields and from other backgrounds (most notably women) to bolster their claims. For example, a careful scrutiny of their thinking and working styles might reveal more similarities than differences between successful male and female scientists. In any case, their arguments would have been more persuasive had they included more women in their sample.

Wolpert and Richards's claim for diversity in approaches to problem selection and problem solving among notable scientists is consistent with the thinking styles of geniuses documented by Michalko (1998) and others (Ochse 1990; Simonton 1988). When asked about how he got the idea of unifying certain parts of the theory on particle physics, theoretical physicist and Nobel laureate Abdus Salam (1926– ) replied:

It's such an "attractive" idea. You see, the whole history of particle physics, or of physics, is one of getting down the number of concepts to as few as possible [*know how to see*]. . . . It always surprises me that some of my physics friends— and some of them very eminent people, Nobel Prize winners—would not subscribe to the idea [*look at the other side*]. They would find the difficulties in uniting two totally disparate looking phenomena so overwhelming that they would think you stupid otherwise [*make novel combinations, connect the unconnected*] [Michalko's strategies are emphasized] (Wolpert and Richards 1988:17–18).

When asked about how he usually went about his work, whether he collaborated or did his work in isolation, theoretical physicist Michael Berry (1941–) answered:

> A bit of each. Sometimes problems come rather naturally and internally as developments from what I've done before. Many of the morphological problems in waves are like that, but sometimes they arise quite randomly as a result of conversations I've had with people [*find what they are not looking for, awaken the collaborative spirit*]. . . . It's really a mixture of keeping my eyes and ears open and developing themes [Michalko's strategies are emphasized] (Wolpert and Richards 1988:44–45).

Based on Michalko's observations as well as sample accounts of outstanding scientists interviewed by Wolpert and Richards, it is difficult to deny a connection between certain thinking styles and successful pioneering work. Individuals have made significant discoveries or innovative ideas in science because they tend to see what no one else is seeing and to think what no one else is thinking. From a cognitive viewpoint, they are indeed pioneers in their fields.

Creative or inventive people are motivated to continuously seek and solve new problems by a desire to excel (Helson 1996). There are two types of thinking behind discoveries and innovations. Divergent thinking is the capacity to produce many alternative responses, both varied and yet highly original, to the same problem. In contrast, convergent thinking entails the capacity to produce a single correct response. Scientific creativity requires more convergent thought than divergent thought, while the opposite is true for artistic creativity (Simonton 1999:85–87). All this means that would-be scientific pioneers must be persistent and tireless in searching for the right answer to a particular problem. Other personal qualities and support mechanisms, however, may be crucial for stimulating and nurturing convergent thought.

## Personal Qualities

Not all creative people have scientific accomplishments. Those who have been highly successful possess certain kinds of attributes that are essential in initiating and pursuing pioneering scientific work. Successful scientists set themselves apart from less successful ones in terms of effort, ability, and dedication to learning and work (Hargens 1978; Moulin 1955; Ramon y Cajal 1999; Simon 1974). Personal reflections in interviews and autobiographies, observations, and anecdotal evidence underscore the importance of other traits such as intellect, drive, ambition, determination, risk-taking, perseverance, and persistence (Hermanowicz 1998). These qualities are also found among individuals who are successful in business or other endeavors (Gardner and Laskin 1995; Harrington and Boardman 1997; O'Connell 2001; O'Donovan-Polten 2001). All this entails the need for independence and autonomy, which is more likely to be a problem for talented women than for talented men (Epstein 1993; Kanter 1977; Valian 1998). This issue will be explored in depth in the next chapter.

John Polkinghorne disputed the claim that geniuses have identifiable personalities or certain preferences for research topics. He observed that two pioneers in physics, Richard Feynman and Murray Gell-Mann, were very different in personality as well as the kinds of work they pursued (1998:46). Despite differences in personalities, scientific pioneers seem to have more commonalities than differences in other tangible qualities. As noted by Polkinghorne and others (Hermanowicz 1998; Zuckerman 1996), these leading contributors to science were recognized early on as exceedingly bright; colleagues or collaborators could not help but feel awed by their scientific talents. Additionally, their research, teaching, or commitment to the professional community stood out in many people's minds. Students were often drawn to them by their research excellence and/or teaching styles. As teachers, they were a constant source of stimulating ideas for students.

Further, Polkinghorne's first-hand accounts of Paul Dirac, Abdus Salam, Murray Gell-Mann, Richard Feynman, and Stephen Hawking corroborate the observations of scientists made by Simmons, Michalko, Wolpert and Richards: What distinguishes scientific pioneers from others is the significance of their discoveries or ideas, creative thinking style, and diverse modes of investigation. What is lacking in Polkinghorne and others' discussions is the role of cultural and social forces in these notable physicists' career development; at different stages of their scientific careers, environmental forces were deemed insignificant. Given that he was able to make observations of how scientific pioneers think and work up-close, it is difficult to understand why Polkinghorne overlooked external influences on the career development of "the best in physics," since many studies underscored the significance of environmental influence on the development of creative or inventive ability (Howe 1999; Ochse 1990; Simonton 1999).

To gain a better understanding of the production of scientific pioneers, we will now examine the background and circumstances of the exceptionally talented.

## EXTERNAL FACTORS

Internal factors comprise only one set of reasons for success or failure in science, and may or may not reflect specific abilities or capabilities in scientific work. Besides individual factors, structural and institutional forces also shape the development and growth of talented individuals.

### Family

Many studies have highlighted the importance of family background and upbringing to the origins and development of individuals with outstanding accomplishments in science, politics, and other domains (Goertzel and Goertzel 1962; Goertzel, Goertzel, and Goertzel 1978; Harrington and Boardman 1997; Roe 1952; Simmons 1996). The environment, family in particular, has a strong, stable, and enduring influence on the development of a person's scientific career. Family literally affects everything that a child does. Also, whether a child attends a public school or private school might facilitate or hamper the development of his or her scientific potential. In her study entitled *The Making of a Scientist*, Roe observes a connection between school types and the production of scientists; scientists generally are more likely to come from progressive schools than from private or church schools (Roe 1952:239–240). So, uncontrolled factors such as family class background and parental decisions on the type of school a child attends might have a lingering, if undetermined impact on the making of scientific pioneers. Regarding family background, many notable scientists were born or raised in affluent homes: professional or economically well-off families. Additionally, there was a keen emphasis on education or intellectual pursuits in their family. Parental influence or encouragement in intellectual/scientific curiosity or pursuits was strong. However, there is mixed support for the concept of familial influence on the development and growth of scientific careers.

To address the question of what makes people creative, Ochse gave an extensive and thorough discussion of familial influence in *Before the Gates of Excellence: The Determinants of Creative Genius* (1990). Her thesis is that creative geniuses are made, not born (e.g., artists, musicians, scientists, writers), and that they appear in certain places under certain circumstances, and are definitely *not* the product of chance (Ochse 1990:50). The shortcoming of Simonton's work is that little attention was given to the social origin and family of the exceptionally talented (1988, 1999). In contrast, Ochse tells us how talents can be turned into skills by looking at the social background and home environment of creative genius.

Would-be creators tend to be brought up in families in which learning is highly valued. Ochse stresses that financial resources are not as important as the encouragement and ambition to excel received from parents. A premium on learning in professional or Jewish families fostered the development of creative ability. The same can be said about children from immigrant families. Equally important, getting ahead in science, for example, requires one to acquire intellectual skill and to be creative. An intellectually stimulating environment is essential for creators-to-be; positive role models from the family, schools, and/or books should be readily available for potential achievers.

There is some indication of birth order effects on scientific innovations. Firstborns are overrepresented among supporters of reactionary ideas (Sulloway 1996:42–43). This is understandable, because firstborns (and only children) are more likely to receive undivided attention from adults, especially parents, whom are the major source of intellectual stimulation as well as material support. Further, compared to laterborns, firstborns are probably under more pressure to excel and to set an example for younger siblings (1996:66–67). Nevertheless, Sulloway has pointed out that laterborns initiated and supported most innovations in science, radical ones in particular. Although firstborns are attracted to reactionary innovations, they are inclined to reject new ideas, especially when these ideas are inconsistent with conventional theories (1996:53–54). To compete for parental attention and affection, laterborns need to carve out a niche in the family for themselves by developing alternative personalities. Being open to novel ideas and new experiences may therefore be a sibling strategy adopted by laterborns.

Based on his analysis of a relatively small sample of women, Sulloway also uncovered several important attributes to distinguish early female pioneers in science. Chief among them was being laterborn. (But for men, being firstborn constituted a strong stimulus to the development of a scientific career.) These women grew up in liberal families and held liberal social attitudes. Unlike men, they were more likely to have conflicts with a parent, usually their mother. As a result, many had a strong identification with the father. They also displayed gender-incongruent behavior, which could be attributed to the prevalence of an instrumental climate in their families. Of course, not all female pioneers exhibited these characteristics. Margaret Mead, whom Sulloway labeled as "a socially liberal firstborn," was not receptive to radical intellectual ideas. Instead of launching a revolution in her field, Mead built on the legacy of her renowned teacher, Franz Boas (Sulloway 1996:157–170). Sulloway attributed Mead's opposition of a liberal scientific revolution to her role as "the protective, dictatorial big sister" to three younger siblings. This authoritarian style was extended to her anthropological work and later refusal to admit mistakes in her best-selling book *Coming of Age in Samoa* (Mead

1928). Troubled homes and bereavement prepared these potential achievers to deal with extreme hardships in life and work constructively. Isolation and loneliness in childhood enabled them to focus and concentrate on intellectually demanding tasks. Authoritarian parental control did not hamper the development of discipline and structure in life and work for these women. Ochse's emphasis on social background and home environment complements Howe's life course approach to unravel the complex process of the production of scientific pioneers.

Discussions in this section underscore the importance of utilizing biographical as well as ethnographic data in research on female scientific pioneers. However, it is imperative to point out that each (auto)biography is a product of the conditions of its creation. An argument can then be made that the data used for this study was highly influenced by our cultural perceptions of how gender functions in science (Fausto-Sterling 1992; Keller 1985; Schiebinger 2001). For instance, there have been questions about the extent to which a(n) (auto)biography or self-presentation accurately portrays a subject's life (e.g., Banner 2003; Comfort 2001; Curie 1937; Keller 1983; Lapsley 1999). Of course, no biographical data is unaffected by cultural changes. Therefore, we should avoid taking these materials too much at face value.

## Cultural/Economic/Political Changes: Good Timing

The fact that less than a dozen women have received *the* universal honor for scientific excellence, the Nobel Prize (Zuckerman 1978:420), says much about the scientific community, and perhaps even more about the society that shapes individuals' attitudes toward science and career outcomes. Whether one rises to the top depends, in large part, on many circumstances beyond that person's control. Scholars have underscored the influence of cultural, economic, and political changes on the operation of scientific establishment. The intertwined relationship between the reward structure of science and historical developments was best captured by the following statement:

> The opening of the Nobel and other archives for research is of great value to historians of science and others, for it allows insights into how the choice of prize winners has been influenced not only by science but also by politics and culture, and provides the opportunity and challenges to reflect on the values underlying yesterday's and today's scientific world (Crawford 1998:1257).

In *Striking the Mother Lode in Science: The Importance of Age, Place, and Time*, Paula Stephan and Sharon Levin (1992) examine the emotional and intellectual climates in which successful scientists are brought up and trained. Results of their analysis emphasize the significance for young or budding sci-

entists of being in the *right place at the right time*; a person would be more
likely to have a successful career in science under three conditions: a superior
vantage point, a stimulating intellectual atmosphere, and a vibrant economy.
Tangible resources such as equipment, supportive colleagues, and chance
have become increasingly important in recent decades. Professional ties and
research support are arguably crucial for developing scientific talents. This is
especially the case when modern scientific research tends to be large-scale
and collaborative.

The science "games" might have changed over the years, yet cultural developments and political changes still have a strong impact on one's scientific career. The interaction of these forces might have directly or indirectly led male and female scientists into different career trajectories.
Simonton has noted that female geniuses are highly unlikely to appear in
certain cultures at certain times. For instance, the Confucian ideology,
which had a prominent influence in the history of Japanese culture, favors
the subordination of women to men in all aspects of life. In Confucianism,
the ideal roles of women include obedient daughters, submissive wives,
and nurturing mothers. Additionally, in Confucian societies, women did
not enjoy the same opportunities as men to receive education or to compete for jobs. Simonton attributes fluctuations in the emergence of female
geniuses in Japan to shifts in the prevalence of Confucian ideology and
other changes (1999:220–221).

Based on the arguments of Simonton, Levin and Stephan, we should consider cultural and structural arrangements in finding out how it is possible
for women to reach the top of the scientific establishment. We should also
attempt to distinguish between short-term and long-term forces that lead to
scientific pioneering work, such as changes in the scientific profession and
society that foster women's participation in science.

## SUMMARY

This chapter provides an up-to-date account on the production of successful
scientists by critically reviewing a number of selected publications. It also
discusses the implications for research on pioneers of modern science.

So, are scientific pioneers born or made? If they are made, as research suggests, how can one become a pioneer in science? Factors underlying the making of scientific pioneers range from family background, creativity, diversity
in approaches to problem selection and solution, and good timing. Preceding
discussions underscore the continuous significance of both predisposing

(cognitive and work habits/style) and precipitating (family background/other influences) factors on scientific success. There is inconclusive evidence as to which factor is more important than the others for career progress. However, the consensus among researchers is that environmental factors can nurture the development and growth of scientific talents, or stifle it.

What can research on scientific pioneers tell us about *women* in science? Preceding discussions are highly critical of recent studies for giving insufficient attention to female scientists. Female scientific pioneers are counterexamples of the "revolving door phenomenon" in science for women: Instead of dropping out after entering science, female pioneers stay in and get ahead in science.

Given the marginal role of women in the development of science, do male and female pioneers "make it" in the same, or different ways? Scientific pioneers (be they men or women) may possess certain persistent traits and characteristics which set them apart from others. Alternatively, are they simply ordinary persons who just happen to be in unique circumstances? There is no lack of research on scientific elites, yet only a few serious attempts have been made to systematically link the topic with high achieving men *and* women in science. Most current studies focus almost exclusively on male scientific pioneers. As a result, current works on scientists with outstanding careers offer an incomplete (and perhaps distorted) account of science makers. We do not know whether and to what extent the conclusions drawn can be applied to female scientific pioneers.

We still lack knowledge about the characteristics as well as the context of the making of female scientific pioneers. In *Before the Gates of Excellence*, Ochse has *briefly* addressed the issue of female creators (1990:172–175). She notes that the background and personality of creative women bear striking resemblance to those of creative men. Like male creators, female creators also come from professional-class homes and are brought up with a strong respect for learning and achievement; they have also observed their parents engaging in intellectual activities, such as reading. Their fathers, though emotionally distant, are the principal figure of intellectual inspiration. Besides these factors, financial and other forms of insecurity are not uncommon in either gender's early experiences. Creative women are naturally shy. As a result, they prefer working alone or being absorbed in solitary activities to interaction: another similarity to their male counterparts. The finding of more similarities than differences between male and female creators suggests that female scientific pioneers might not be very different from their male counterparts in terms of circumstances and other characteristics.

Yet women continue to be significantly underrepresented in science, particularly among scientific pioneers. It is true that most of the formal obstacles
women face have been dismantled. However, as shown in the forthcoming
chapter, informal barriers, hard to identify, continue to exist. By studying female scientific pioneers, perhaps we can see these underlying forces more
clearly.

*Chapter Two*

# Why Aren't There More Female Scientific Pioneers?

Why is it that nearly one hundred years after Marie Curie shared her first Nobel Prize with Henri Becquerel and Pierre Curie, there are so few prominent women in science and medicine? [quote from *Rosalyn Yalow: Nobel Laureate: Her Life and Work in Medicine* (Straus 1998:xii)]

## INTRODUCTION

The simple answer to the above question is: fewer women than men enter science training, and there are also proportionally more women than men leaving science. However, the "leaky science pipeline" thesis is not an adequate explanation for the virtual absence of women among scientific pioneers. According to Charles Darwin, for example, his sister was brighter than he was. When she was a child, Adele Galton tutored her brother Francis Galton (Ochse 1990:172). There is no shortage of theories regarding the differential success of men and women in science. Explanations for a persistent gender gap in science representation range from differences between men and women in intellectual abilities and upbringing, to science culture, to discrimination. As seen below, each explanation sheds a different light on the issue of women's underrepresentation in science.

## EXPLANATIONS OF SCIENCE GENDER GAP

To set the stage for analyzing the origins and development of female scientific pioneers, we focus on four major arguments that contend with and/or complement one another: (a) nature, (b) nurture, (c) social control, and (d) environment. How useful are these theories in explaining gender inequality in scientific achievements? To what extent can we apply each approach to understand the career development and career advancement of female scientific pioneers? Besides reviewing their strengths and limitations, when appropriate, I will highlight major differences, or draw parallels, between certain approaches. I also discuss how each of these approaches predicts the career achievements of female scientists. Throughout the chapter, I also present relevant issues that will be addressed in subsequent chapters.

## The Nature Argument: The Biological Approach

According to proponents of the biological approach, men and women have different rates of participating in science because of innate differences in abilities and talents: that both physically and emotionally, women are ill suited for doing science (Fausto-Sterling 1992; Gieryn 1995:422–423; Schiebinger 2001).

The nature of science requires certain attributes from its practitioners. In addition to having a passion for science as well as an inquisitive mind, scientists are expected to be objective, detached, and have a risk-taker's aptitude. It has been found that innovative ideas or discoveries usually come from those who are determined, self-confident, and persistent, and that to become a successful scientist, one must have the capacity to focus and concentrate on the tasks at hand, to devote themselves to work, and to be competitive and aggressive (Ramon y Cajal 1999). Others have observed that single-minded dedication to work is the most salient characteristic among creative people. This is why they have a tendency to work alone, and to do whatever they can to avoid disruption (Ochse 1991:338). These qualities for the making of a scientist, according to proponents of the biological approach, are seldom found among women. In contrast, women are generally perceived as emotional, compassionate, sociable and less competitive, all factors which are related to women's reproductive capacity and domestic roles. Some scholars also invoke the argument of other innate differences, such as size of the brain, to explain why women are unfit to pursue careers in science.

What makes the "nature argument" distinct is that it allows any observed gender inequality in science to be attributed to women's deficits in abilities, talent, and other traits necessary for success in science. With this argument, neither the individual nor the society is held responsible for the virtual absence of women among scientific pioneers. This "blaming mother nature" approach would appeal to policymakers—if accepted, very little intervention (if any) from any party can be done to rectify the persistent gender gap in science. It also implies that individuals can initiate little or no action at all (even if they want to) to make up for any intellectual and personality deficits.

Despite its persuasiveness to many, the biological approach has many drawbacks. It has limited power in explaining women's relatively low level of entry and participation in science. First, its assumption of women's intellectual deficits in relation to science is inconsistent with the historical evidence of (a) women who made scientific discoveries or innovations (Rayner-Canham and Rayner-Canham 1997; Schiebinger 1987:308–315; Stanley 1995), and (b) women who successfully managed science projects or ran research labs efficiently by themselves or along with men (Kohlstedt 1999; Mozans 1991).

Second, the genderization of science—equating science with male attributes such as objectivity and reason—has been challenged as an ideology to justify sexual division of labor in society, and perpetuates gender differences in career achievements (Jacobs 1995; Tomaskovic-Devey 1993; Williams 1995). Genderization's function is to make science inaccessible to women and preserve male domination in science (Keller 1985). The prevalence of this ideology may lead to different treatment of male and female students in science and mathematics classes. In fact, there have been observations of science teachers giving more support to male students than to female students. This differing teaching style could result in a loss of interest and self-confidence rather than lack of abilities to learn and pursue science among female students (NSF 1994:21).

Third, the biological approach contradicts the finding of comparable levels of completion of science and mathematics courses among male and female high school students. Gender differences in science performance begin to emerge among twelfth graders (NSF 1999:12–15). Even if there are innate differences in scientific talents between men and women, according to the biological approach, somehow they are suppressed until midchildhood.

Fourth, this conservative approach fails to explain why female college students tend to leave science majors in droves after their freshman year (Seymour and Hewitt 1997). The "revolving door" phenomenon observed among female college students suggests reasons beyond a lack of abilities.

Fifth, the assumption of biological determinism is further weakened in light of the success of various school programs conducted outside traditional laboratories in sustaining comparable levels of interest in science between male and female students (Eisenhart and Finkel 1998).

A one-dimensional view of gender inequality, the biological approach offers a simplistic and limited account of gender differences in the entry and participation of science. In this outlook, individuals are seen as passive agents of change. Science is a social activity in many respects, yet this approach fails to consider other factors. What roles do family, schools, and society play in shaping individual decisions and actions? To unravel the complex process of

the production of scientific pioneers, we need to look beyond this static view
of human actions.

## The Nurture Argument: The Individual Choice Approach

The individual choice approach offers a dynamic view of human interactions.
Career development and advancement are outcomes of a convergence of
choice, design, and adaptation (Evetts 1996; Harrington and Boardman 1997;
O'Dononan-Polten 2001). In this view, people constantly make conscious decisions on career progress in light of their own interest and circumstances; individuals are active agents in choosing between opportunities and constraints.

Nurturing and individual choice are influenced by external factors. Images
of science and scientists have shaped the attitudes of parents, teachers, and
students toward science. Generally, science has been portrayed in textbooks,
as well as in the media, as a male domain (LaFollette 1988; Richardson and
Sutton 1993). Because of its intellectual demands, practitioners are expected
to be devoted to and committed to their work to the exclusion of their personal lives. These images conform to the social expectations of men. In reviewing the developmental history of female intellectuals, Maccoby
(1970:22–23) observed that at some point in their childhood, they were
"tomboys." She argued that development of high intelligence might be associated with cross-sex typing: that people with high IQs have more of the interests and activities characteristic of the opposite sex. For example, boys
with high IQs have more feminine traits, and girls with high IQs tend to have
more masculine traits, including greater independence and dominance and
less submission to authority. Maccoby and others have added that unlike men,
women have a motive to avoid success (academically or professionally),
since girls who are too successful would be unpopular among their peers and,
more importantly, become unattractive to men. Indeed, it is deemed impossible by some for a woman to be deeply interested in and seriously engaged in
intellectual pursuits. In this line of thought, scientific career endeavors are incompatible with marriage and family obligations, and women who choose
science as a career are often regarded as unfeminine and deviants (Morgan
1992). This is probably why those who succeed in reaching the top in male
endeavors run the risk of being labeled "superwomen" or "tokens" (Kanter
1977). Women aspiring for science careers are also confronted with this
dilemma.

The gender-role socialization thesis has been used to account for gender
differences in career orientations (Jacobs 1989; Sonnert and Holton 1995).
This thesis suggests that, because of their traditional upbringing, men and
women are inclined to choose careers corresponding to their gender roles.

One can make the argument that women's underrepresentation in science as well as their notable absence among scientific elites is a reflection of their desire to meet gender-role expectations, and the fact that few women succeed in science is a manifestation of their rejection of images of scientists or the demands of a full-time scientist's job. The good news is that, according to Barber (1995), changes in the image of women as scientists have taken place since the 1960s. Social transformations, for example, have reduced the negative images of female scientists. As a result, the number of women who choose to study science has increased.

Unlike the biological approach, the individual choice approach places emphasis on creativity and variation rather than uniformity in each person's orientation to career development. Contrary to common perceptions, there are no uniform predictors of career success or failure in science. A positive "kick" (opportunity) may boost the career of a young scientist, but might have no lasting impact on another scientist's career. Conversely, a negative kick (constraint, setback) may deter one scientist from surging ahead, but the same impediment might encourage another scientist to persevere. In effect, the same factor or situation could have opposite effects on the career development of two equally talented scientists. The outcome depends on how a person perceives and reacts to the kick (Etzkowitz, Kemelgor, and Uzzi 2000). This is a corollary to a situation many professional women face: Due to competing commitments to family and work, educated women might avoid positions with rigid career lines and prefer less-demanding jobs. This is possibly why female scientists are disproportionately represented in temporary jobs such as postdoctoral or nontenure-track teaching appointments (Blair-Loy 2003; Reskin 1976:610).

The individual choice approach challenges the assumption of biological determinism. Instead of focusing on personal abilities as the primary cause of success or failure, it underscores the interaction between personal and societal factors. This dynamic, multidimensional view understands career development in the context of the reciprocal relationship between individual actions and culture. Compared to the biological approach, the individual choice approach contributes more to our understanding of the science gender gap. However, there are also weaknesses in this sociopsychological explanation.

First, as active agents of change, individuals are expected to initiate actions to set and change the course of career development in spite of or because of socialization. The biological approach underestimates environmental influence on career outcomes. The individual choice approach overestimates the impact of individual control on career outcomes.

Because of differential access to resources and support, how realistically can we expect individuals from different backgrounds to resist larger social forces (e.g., social class and religion)? Like the earlier example of classroom treatment, how can female students retain and perhaps strengthen their self-confidence and interest in science in light of teachers' different expectations? Even school counselors might encourage male students to pursue science careers, but advise equally talented female students to choose careers in nursing or teaching. Based on this knowledge, how can one assume that the disproportionate number of women departing from science training or employment is a direct result of conscious personal decisions? Chakravarthy, Chawla, and Mehta (1988:58) have contended that devotion to and creativity in science is a combination of individual competence and environmental support. Opportunities to engage in intellectually challenging and stimulating activities, for instance, are essential for the development of professional identity as well as an inner sense of competence. All this can lead to greater scientific productivity. Conversely, blocked opportunity not only hampers a person's chance to gain new knowledge and experience, but also deprives that person of a chance to enhance his or her status and esteem. Thus, based on Chakravarthy et al.'s argument (1988), a person's career choice is a product of perceived extrinsic awards and intrinsic satisfaction. This helps explain why a large number of women self-select out of science by "discouraging" themselves from continuing further study in science (Barber 1995; Schiebinger 1987:321).

Second, the nurture argument fails to address the historical evidence of formal and informal discrimination against women's participation in science. Scholars in the history of science as well as sociology of science have chronicled numerous obstacles women face when they enter and advance in science (Tang 2003). Changing one's style of doing research or postponing marriage or family plans is a conciliatory approach to achieve career advancement. These strategies might help some individuals to make it to the top of the science field, but they do not change the practices and norms in the society and scientific community. Individual actions might be effective in the short run for some female scientists; however, these individual efforts seldom have any long-lasting impact for those who come after the actors. The individual choice approach cannot tell us if (and how) systematic changes introduced collectively by policymakers and gatekeepers of science have affected women's participation in science training and employment. Even if a young female scientist does everything right, there is no guarantee she will make it to the top of the science field.

Third, it is difficult to deny the contribution of historical developments to women's increasing participation in science. The nurture argument overlooks changes in society that constantly shape an individual's career decisions (e.g., industrialization, the women's movement, and affirmative action policies). We have probably witnessed the most dramatic increase in women's participation in science in recent decades due to the introduction of mentoring, fellowship, and sponsorship programs, among others. Additionally, a rising demand for scientific personnel in western societies during the last half of the twentieth century has provided incentives to schools and employers to initiate efforts to make science learning and science careers attractive to young women. Critics of the individual choice approach could argue that substantive collective actions are needed to further reduce gender inequality in science. For instance, to facilitate the production of female scientific pioneers, there should be more intervention on the part of the gatekeepers of science and policymakers.

## The Social Control Argument: The Structural Approach

The social control argument complements the nature and nurture views of the science gender gap. Structural theorists underscore the impact of external influences on career experiences and career outcomes (Kalleberg and Sorensen 1979; Kerckhoff 1996; Osterman 1996). Among those influences, the choice of a field of study or career path is a product of structural arrangements. Lack of scientific role models for women and negative early school experiences in science can discourage women from entering science as well. Similarly, lack of support and resources to pursue scientific research can result in early departure or blocked mobility.

Gender is embedded in educational and career processes, as reflected in cultural, economic, and political arrangements, and cultural bias and stereotypes contribute to the production of relatively few female scientific pioneers. Family and social upbringing have already cast men and women into different roles in the society. Generally, members of society have different expectations of men and women — men are considered doers, creators, discoverers, or defenders who engage in speculation and invention, whereas women are responsible for ordering, arrangement, and decisions at home. For instance, parents generally raise their children differently according to gender (Simonton 1999:219) — competitive/dominant roles for men and nurturing/supportive roles for women. The pervasiveness of these images of men and women helps explain Marie Curie's predicament when she took over her husband Pierre Curie's class after his tragic death. Seeing a widowed woman in an unfamiliar, authoritative role made many of her students uncomfortable. For example, French students ridiculed her Polish accent and rigid style (Curie 1937).

Even if women choose to pursue science careers, they are more likely to be found in social science fields (soft/compassionate science) than in natural or physical science fields (hard/dispassionate science). The opposite is true for men. Due to personal and professional interests, females and minorities are more interested in studying social issues related to gender or race (Kulis and Miller 1988). However, uneven field distribution by gender is insufficient to explain why there are still so few women joining the ranks of scientific pioneers. This phenomenon might be explained by differential access to resources. Compared to soft science research, hard science research is more expensive. To test new ideas, make observations, or conduct experiments, hard science requires many institutional resources. Research funds, facilities, personnel, and collaborators are essential for start-up and expansion. By comparison, soft science requires fewer resources to test new ideas or theories (Nowotny 1991:153).

Cultural influence is reflected in the organizational dynamics of workplaces (Berg and Kalleberg 2001; Powell and Graves 2003), so gender differences in career mobility might be attributed to disparity in work activities. Women in professional occupations tend to occupy important but subordinate roles. Job segregation by sex is an extension of sex roles. For example, women in science are more likely than their male counterparts to perform typically female work (e.g., teaching or research assistants/associates) or are concentrated in women's fields (e.g., home economics, psychology or psychoanalysis) (Chodorow 1991). Female scientists are likely to do "women's work" or to hold a "women's position" such as dean of women (Nidiffer 2000). By comparison, men are inclined to perform high-prestige, masculine activities (e.g., leadership positions) or are employed by research-oriented universities.

Money costs might also be a consideration; women are concentrated in low prestige, part-time positions because employers want to minimize labor costs. Tomaskovic-Devey, Kalleberg, and Marsden have observed that employers may design division-of-labor that relies heavily on women or part-time workers because these workers are often paid comparatively low wages (1996:294). The same analogy may apply in explaining the historical concentration of women in junior or women's colleges. Consequently, when compared to men, women are more likely to hold jobs with little or few prospects for promotion and are less likely to hold decision-making positions. Chakravarthy et al. (1988:52–57) also observed a hierarchical distribution in the division of the scientific labor force. Female scientists were less likely to participate in administration or management. In addition to locating in the lower rung of organizational hierarchy, few women were part of the informal professional networks. This thesis suggests that women have little chance of joining the ranks of scientific pioneers.

Another factor might be that career mobility is mediated by networks. The task of job searches and career advancement require additional information, preparation, and assistance (Granovetter 1995). A lack of established ties is a serious obstacle for women to getting ahead in professional occupations. The societal norm of homosociality tells us why women have difficulties in gaining entry into the *old-boy network*—a form of social capital essential for career development and advancement (Lipman-Blumen 1976; Rose 1989:353). Aside from similarity in background and all other things being equal, men as well as women have more to gain professionally by associating with men rather than women. Because of their *master* status in the society, men generally have a greater control over a wider range of resources. By comparison, due to women's subordinate role in the society and limited access to resources, they are in a weaker position to offer professional assistance and support to others. (In fact, men who associate with women as equals run the risk of being marginalized.) Thus, as gatekeepers (controllers) of resources for career development and advancement, male networks are the preferred choice of social interaction and professional collaboration.

Female networks are less useful. For one, women generally control fewer and less important resources in the society. This can be attributed to their not being part of a much larger pattern of a male homosocial world. Women's lack of resources and crucial information makes the *new-girl network* a less valuable asset relative to the old-boy network. Further, because male-initiated actions tend to be taken more seriously than those initiated by women, most women would have a relatively hard time building up their reputation in the scientific community (Evetts 1996). All this suggests that female students and scientists may have more difficulties in forming important professional ties or finding sponsors, role models, and mentors. This professional isolation would have an adverse effect on reputation-establishment for female scientists. My estimation is that female scientific pioneers are inclined to ally with men and to become integral parts of the old-boy network. However, following from the "more like us" argument, why, then, would influential male scientists bother to mentor or collaborate with women? We will explore this issue extensively in Chapter 6.

There is plenty of evidence to support the social control argument (Rossiter 2003; Schiebinger 1987; Zuckerman, Cole, and Bruer 1991). Many contributions to science made by women are deemed less important and have gone unrecognized publicly. Women were virtually absent in playing any significant roles in the development of science, but not because of a lack of knowledge or expertise. Many female discoverers or inventors remained in the background because of cultural, economic, and political arrangements. The history of science is replete with evidence of legal barriers against women's participation in science. The centuries-long exclusion of women from receiving formal education is a case in point. Apprenticeship and self-teaching became the principal methods of training women. The historically invisible role of women in science reflects and reinforces the ideology of job segregation by gender. Any formal support or public recognition of women's contributions would run counter to these cultural beliefs, which in turn would challenge male dominance in society and in science.

Based on the social control argument, how is it possible for women to break into male domains? Are there really no counterforces to the barriers women face in science? The answer lies in the cultural, economic, and political arrangements of society. At some point, the scientific community opened its door to women, and a few exceptional women could join the ranks of scientific pioneers. Lipman-Blumen refers to this as the "Marie Curie phenomenon" (1976:22). These successful outcomes, however, should not be construed as the elimination of barriers against women's participation in science. The First World War and industrialization, however, gave women viable alternatives to move out of family oriented or traditionally female jobs into a wider range of occupations. During the first half of the twentieth century, political and economic developments in Europe and the United States generated new job opportunities for women. Industrialization and development in defense and information technology industries created a rising demand for scientific and technical workers. Employers turned to nontraditional workers, such as women, who constituted a convenient reserved labor pool. Many women responded to these new market forces enthusiastically.

Additionally, political changes, such as affirmative action programs, have also increased opportunities for women and minorities in education and employment. The essence of affirmative action policies is to redress past grievances by imposing legislation on schools and employers. These institutions are prohibited from discriminating against applicants in admission and hiring decisions based on functionally irrelevant characteristics such as gender and racial/ethnic background. This legislation, along with other developments including the women's movement of the 1970s, has lowered the barriers considerably for women to enter and participate in science. The changing roles of women in science during the last century lend credence to the social control argument that labor market arrangements dictate the timing and number of workers entering a particular field.

If women have not been a part of the process that allocates critical resources and recognition, how have female scientists joined the ranks of pioneers in their field? My estimation is that women capable of pursing pioneering work have received enthusiastic support and guidance from powerful male scientists, since "climbing up the ladder" in many occupations and organizations is most likely "by invitation only" (Epstein 1970:969). This is also true when women become gatekeepers in science, such as advisory board members, journal editors, and grant proposal reviewers. At the early stages of one's career, "what you know" regulates the entry and placement of new hires. All else being equal, as a person tries to get ahead, "who you know" as well as other subjective criteria may take precedence.

Men are not necessarily the only beneficiaries of the old-boy network and the practice of homosocial reproduction in the scientific community. Women, too, have benefited from both the old-boy and old-girl networks in sponsorship, hiring, and promotion. Subjects in two recent studies of female mathematicians reported receiving significant support from male mentors and male role models (Henrion 1997). Traditionally, there have been scant job opportunities for women with a doctorate in mathematics at major universities, coeducational, and liberal arts colleges, so men were more likely to be recommended and hired for openings at these institutions. By comparison, there were more job openings for women Ph.D.s at women's colleges.

So, what prompted a departure from the prevailing norm of hiring male over female mathematicians at academic institutions? During the midtwentieth century, the older generation of women on the faculty of women's colleges hired young women with Ph.D.s who seemed to fit the personal and professional expectations of the older generation. These new hires eventually became a new generation of leaders at these women's institutions (Murray 2000:155–156). Thus, the hiring practices of female mathematicians at women's colleges support the claim of homosocial reproduction in the scientific community.

While worthwhile to consider, the structural approach gives us only a partial understanding of why women have difficulties in joining the ranks of scientific pioneers. There is a wide range of external forces contributing to job segregation by gender, including women's high exit-rates from science training and employment, as well as their tendency to cluster in short-term, lower-tier positions. Nevertheless, structuralists have overlooked the positive aspects of cultural, economic, and political arrangements on women's career attainment and mobility.

## The Environmental Argument: The Institutional Approach

The environmental argument complements previous explanations of gender inequity in science. For the most part, career choices reflect a combination of "push" and "pull" factors. Women are less inclined than men to enter science because of inadequate encouragement and institutional support along with having a strong orientation toward people-oriented careers. There is a decline in women's involvement in science at each of their career-stages due to (a) the pull of marriage and family life and (b) the push of discouragement in graduate school, lack of mentors, and limited career prospects (Preston 2004;

Seymour and Hewitt 1997). Additionally, women's "late arrival and early departure" from science may reflect their experience and treatment in schools, the workplace, and the profession as a whole. Female scientists not only have to cope with the conflict between professional and personal identities, they also have to meet the demands of climbing the hierarchy of scientific establishment.

For many, family is perhaps the first and most important agent of socialization (Bengtson, Biblarz, and Roberts 2002). Generally, unlike marriage or career, family is not an institution of one's choosing. Through encouragement and support, however, the family can foster individual development and growth. Among other benefits, it can help positively shape one's career aspirations and outlook on life. As shown in the previous chapter, a supportive family environment is no anomaly in the production of notable scientists. Research of the eminent in different fields has examined the association of family dynamics with the cognitive and social development of children.

The evidence for a positive impact of family origin on careers is inconclusive, indicating that being born and brought up in a middle-class or professional background does not necessarily produce a scientific genius or political heavyweight. Yet most notable figures had (grand)parents who were highly educated or had professional or business experience, and learning and serving were ways of life in their family tradition (Ochse 1991:337). Maccoby notes that female mathematicians reported having close relationships with their father in childhood and identifying strongly with their father rather than with their mother (1970:22). Studies on "creative" female mathematicians conducted by Helson (1971) also showed their strong identification with the father.

Even if people do not follow the example of their parents to enter the same field or profession, or help run the family business, many of them may tap into the family resources for personal and professional development. Has this been the case for both male and female scientists? How does the family affect the development of female scientific pioneers, who invariably have overcome enormous barriers? We could learn more about the significance of family on career development and advancement by analyzing the backgrounds of women with outstanding career achievements. Unfortunately, the bulk of studies on famous people focused almost exclusively on men.

What role has family played in the production of female scientific pioneers? My belief is that parents of female scientific pioneers tend to provide an intellectually stimulating home environment; there would be parental actions that foster the development of curiosity and independent thinking in their children. Specifically, they tolerate or even encourage their daughters to develop masculine traits, and to engage in activities traditionally thought of as male. Thus, among female scientific pioneers in their childhoods, we should expect to observe "tomboy" behavior, "fighting back," or exploration (Maccoby 1970:21–25). Also, has the family been a source of unconditional support for women with an interest in science, or of discouragement and frustration? Or, could the parents have played conflicting roles? That is, one parent encouraged the daughter to pursue a career in science, while the other steered her into other careers? It has been suggested that, unlike boys, girls are not expected to engage in serious intellectual pursuits. Even if they do, these intellectual interests are simply hobbies rather than preparation for careers. This is one reason why women were historically denied access to education (Ochse 1991:337–338). For female scientists who had very supportive parents, how and why did they help their daughters enter what is considered gender-inappropriate careers? Conversely, for those who did not have parental blessing or approval, how did they overcome the first major obstacle to entering and advancing in science?

Not only that family upbringing matters, but also that young women aspiring for science careers can continuously rely on this support mechanism throughout their academic and career development. For instance, studies have shown that marriage and parenthood have a positive impact on men's careers, while the opposite is true for women. The cause could be that, due to women's expected roles as wives and mothers, employers and sponsors might be reluctant to invest in the training of women (Epstein 1970:969–970; Lipman-Blumen 1976:20). Another possible cause is that, to stay in or get ahead in science, women, like men, are expected to put work and career above personal life. As noted in earlier sections of this chapter, marriage and children mean different things for men and women. Women are more likely than men to perceive work and family conflicts as serious problems. This is understandable, because men can rely upon their wives (and domestic help) to shoulder the burden of household and childrearing responsibilities, while women are ordinarily expected to meet both the gender and family roles.

Regardless of the underlying factors, marriage and children can pose tremendous challenges to women in professional occupations (Blair-Loy 2003). This argument is corroborated by Sonnert and Holton (1995), who observed that most women in academic science, for example, faced the unique challenge of synchronizing three clocks. Unless they chose to be single, postpone marriage, or avoid parenthood, married women had to beat their career clock (meeting the requirements for tenure within a specific period of time), biological clock (childbirth), and career clock of their spouse (likelihood of geographical mobility and career disruption). During the most productive period of life (e.g., 20s, 30s, and 40s), women were most likely to be burdened by maternity and childcare responsibilities. Even with domestic help, women were seldom able to completely devote themselves to work, and expected to be regularly interrupted by others. For female scientists, constant demand for attention was a fact of life. In contrast, male scientists often had wives to take care of nonwork-related demands (Ochse 1991:339–340).

Female scientists in industrial settings face similar challenges (Evetts 1996). In a recent study of U.S. scientists and engineers, Xie and Shauman (2003) concluded that the gender gap in parenting responsibilities is a critical barrier to women's advancement in scientific and engineering professions. The question then becomes whether marriage and parenthood present similar challenges for female scientific pioneers. Do most of them strongly avoid family and parenting roles? For those who are married, how do they handle the demands of family and work? It is clear that female scientists with outstanding careers are a disproportionately small group when compared to their male peers. As a group, female scientific pioneers have set themselves apart from other scientists, including male and female scientists, in terms of characteristics and performance. Aside from differences in living style, female scientists with outstanding achievements may also differ from others in terms of determination as well as ability to find ways to juggle between family and work (White 1982:955).

Schools, as training grounds as well as employers of a large number of scientists, have gate-keeping functions in the production of female scientific pioneers. The impact of teachers' style on student performance has been addressed in earlier sections. Receiving a degree from a prestigious department or studying with an eminent mentor also improves a scientist's chance of getting a prestigious academic job (Long, Allison, and McGinnis 1993:705). Many Nobel laureates have also studied under great masters (Zuckerman 1996).

Men's accumulative advantage over women emerges at schools, however. Apart from differential classroom treatment, overall dynamics in colleges and graduate schools are responsible for the persistently high attrition rate from science among female students. It is true that comparable amounts of resources are available to both male and female students at the universities these days, yet there is a persistent gap in allocating these scarce resources to men and women. Scholars have observed gender differences in access to critical support for formal and informal education, such as scholarships, fellowships, role models, and mentors. In addition, they contend that many practices in formal training, sponsorship, distribution of federal and private funding, and hiring favor men over women in science (Schiebinger 1987:320; Tsoi 1975:111). In a study of academic scientists, Long, Allison, and McGinnis (1993:719–720) found that the process of rank advancement is affected by particularistic factors. Compared to men, women take longer to gain promotion to higher academic ranks. Women also have to meet higher standards for promotions. Additionally, being in a prestigious department has a negative effect on promotion for women. If this is the case, how do female scientific pioneers circumvent these barriers?

There is some indication that women tend to perform better in a supportive work environment. As Chakravarthy, Chawla, and Mehta have argued, work groups and colleagues constitute powerful forces in shaping a worker's attitudes and behavior (1988:70). Further, institutional settings and other social mechanisms, which have a strong bearing on professional identity and commitment, can foster or impede one's career development (Powell and Graves 2003; Reskin 1998). As the center of work life and collegial interaction, a department's culture has a bearing on scientific careers. Female scientists tend to perform and fare better occupationally in "relational departments" than in "instrumental departments." Relational departments ("good departments") have a positive impact on women's careers, because a relational department's culture is characterized by cooperation and collegiality, and this fosters collegial interaction and communication. The atmosphere is favorable to the professional growth of female faculty. In contrast, instrumental departments ("bad departments") have a negative impact on women's careers, because competition and hierarchy characterize the instrumental department's culture. This discourages collegial interaction and fosters professional isolation among female faculty, so is detrimental to the career development of female scientists (Chodorow 1991:176; Etzkowitz, Kemelgor, and Uzzi 2000:179–186).

Do female faculty members perform and fare better in a department when they constitute a critical mass (i.e., ≥ 15%)? It is quite possible that when the proportion of women reaches a critical mass, male behavior toward women might change and the number of female role models available to women may increase. As a group, women in these departments could mobilize collective actions against discriminatory practices and harassment. Yet the effect of "tokenism" in the workplace is not gender-blind (Yoder 1994; Zimmer 1988). Etzkowitz, Kemelgor, and Uzzi (2000) observed that an increase in the number of women in a department alone, without any changes in its culture, would not bring about gender equity in career achievements and advancement. For one, despite an increase in their number, if women in a department are scattered in different fields as subgroups, sex segregation by field could still perpetuate their self-isolation.

Double standards for male and female behavior in the workplace present another impediment to the possible advantage of critical mass to female faculty's career development. Others have pointed out that the parameters for acceptable professional conduct and behavior are broader for male than for female scientists in academic or industrial settings (Shapley 1975). Female scientists are often considered by their peers a woman first and a scientist second. This is generally not the case for male scientists (Yentsch and Sindermann 1992). Thus, if a department's instrumental culture remains unchanged, it is highly unlikely that a critical mass of women in a department will improve the situation for female students and faculty.

All this suggests that only those in positions of power can bring about any substantive changes in a department, and that female students and faculty enjoy strength in numbers only after getting support from key decision makers in a department. These power brokers, who tend to be males, can serve as catalysts for change in the workplace (Simonton 1999:220). A department head, for example, can be a strong advocate for junior female faculty. As someone with strategic power and resources, department chairpersons can support a relational departmental culture to facilitate collaboration and collegiality among colleagues from different backgrounds. Direct support from gatekeepers can be the most critical element in bringing about institutional changes.

Like many professional occupations, science is a demanding institution in terms of its expectations of allegiance and commitment from practitioners. The same argument has been made about the demands of family on women in science (Nowotny 1991:156). But unlike some of the venerable professions, such as law and medicine, science in general has low barriers to entry. For this reason, science is a profession conducive to women's entry (Blalock 1967:92–97; Reskin 1976:597). History is replete with examples of individuals who made scientific discoveries without formal training. Women and racial minorities were historically barred from formal participation in the development of science, yet their marginal roles did not prevent them from doing scientific research in the background and making contributions. Slowly and surely, the formal barriers for their entry into science training and employment were lifted.

In spite of lowered barriers, there is an opportunity-cost of staying in science for women, personally and professionally. Scholars have singled out the resistance of cultural and organizational changes in science as major impediments to women's careers (Etzkowitz, Kemelgor, and Uzzi 2000), because women are expected to adapt to the prevailing male culture of science (Barber 1995; Gieryn 1995; Shapley 1975). However, this male-dominated culture has made it difficult for women to succeed. Compared to men, women are less likely to have technical or scientific "tinkering" experience in their upbringings (Lonsdale 1970:57; McIlwee and Robinson 1992). As a result, women in science tend to have more difficulties than men do in forging ties with their male peers for career advancement. Numerous studies have also suggested that it is somewhat problematic for women to fit into male-oriented work cultures (Chetkovich 1997; Epstein 1993; Swerdlow 1998; Williams 1989). Whether perceived or real, this may adversely affect women's job performances, and subsequently their career outcomes. This is probably one reason why women have a somewhat high rate of attrition from male-dominated professions, such as science, engineering, and computer science (Margolis and Fisher 2002; Preston 2004; Wright 1997). These high attrition rates may very well be the outcome of their dissatisfaction from functioning in male-oriented work cultures. This "square peg in a round hole" perspective also tells us why graduate schools and employers have failed to retain female students and scientists.

Then what sets the successful women in science apart from the less successful ones? The ability to fit into the culture of the science-related workplace is one factor. Finding scientific work and organizational practices appealing is another. To get ahead in any work setting, successful women might have adopted a set of strategies to deal with challenges and setbacks. It is possible that some female scientific pioneers adapted to and assimilated into the existing cultural norms of science and the science-related work culture, while others attempted to transform them.

The environmental argument addresses the important question of why female scientists still have a difficult time excelling in science relative to men. It is true that historical developments have dismantled formal barriers for women in science training and employment. But these changes would not and could not completely change the norms and practices of other institutions. Upon entry to college and labor markets, the forces that draw women away from science can come from their families, schools/workplaces, and the profession itself. Many of these constraints are beyond their control. Unfortunately, the structure of the scientific profession is based on competition and research productivity. In particular, the conventional measure of success in the scientific-academic community, research productivity, tends to favor those who follow a linear career path with few or no interruptions.

This approach, however, puts too much emphasis on a person's willingness to subscribe to the traditional scientific role model, which generally requires a scientist to (a) take on an aggressive persona; (b) follow a somewhat rigid, linear career path; and (c) commit full-time to work. These expectations are common to many professional occupations (Barker 1999; Chetkovich 1997; Epstein 1993). However, these requirements are particularly difficult for female students and scientists to meet.

Supporters of this model have certainly overlooked the counterforce to the hierarchical structure and practices in science. During the past century, women, along with members of racial and ethnic groups, constituted a new generation of scientists. Despite their recent entry and small group size, they might serve as a catalyst of gradual change through conformity with as well as resistance to external pressures. In addition to *following* the traditional linear career path, there is a possibility of *creating* alternative career models among both male and female scientists. Indeed, for recent generations of scientists, dealing with the conflicting demands of family and work may help loosen some of the institutional constraints. For example, a new generation of male scientists may help facilitate cultural and structural changes in a department and at home (Etzkowitz, Kemelgor, and Uzzi 2000). They, too, might want or need to balance work and family demands. Like female scientists, male scientists may adopt a collaborative/proactive style in their work and peer interactions. They can become strong advocates for female scientists and potential allies with their female colleagues. As spouses, they can provide critical support to their scientific wives, domestically and professionally.

In short, the institutional explanation underestimates the likelihood that a new generation of male and female scientists, replacing the generation of older male scientists, makes the development of alternative scientific role models possible.

## SUMMARY

Each of the arguments discussed in this chapter has direct implications for career development and career advancement in science. The *biological* approach underscores the influence of ascribed characteristics on educational and career outcomes. The *individual* approach revolves around personal interest and preference for certain academic and career choices. Both the *structural* and *institutional* approaches attribute career success or failure to external forces, whether opportunities or constraints. The structural approach highlights the interaction of cultural, economic, and political developments in shaping career prospects and processes. The institutional approach focuses on the differential impact of family, school, and profession on career advancement. In the case of scientists, career attainments and mobility are dictated by transformations in society as well as changes in the scientific establishment. One cannot discuss the constraints of one particular institution without considering the others.

The overview explanations of science gender gaps raise an important question about the relative influence of different factors on career success. We should view success in science as a product of multiple forces—individual, structural, and institutional. This observation is consistent with the argument of Chakravarthy, Chawla, and Mehta (1988:72), who emphasized the need to examine impediments to women's career advancement at three levels: (a) the personal level—women's own inhibitions, (b) the group level—the lack of encouragement, and (c) the societal level—the lack of recognition.

A review of the literature on the science gender gap reveals several important themes. Not all individuals with scientific talents enter into or remain in science. Personal preference, cultural ideologies, and structural as well as institutional constraints might have made science a less attractive career option for women. The relatively low level of women's representation in science may also reflect their frustrations with the scientific professions. These are not encouraging signs for young women aspiring to science careers.

Our conventional focus on the educational and occupational barriers for women does not convey the course of career development of women reaching the top. Despite differential access to resources, career decisions and career actions of male and female scientists, there are compromises between talents, choices, and constraints. As the number of women participating in professional occupations increases, it is expected that fewer female scientists will give up their pursuit of career success, and many of them will find new ways to juggle between family and work despite cultural and structural constraints (Glover 2000; Pattatucci 1998).

This chapter has provided a general prehistory of female science makers. In the next three chapters, I analyze the life, work, and career histories of ten notable female scientists. While coming from different backgrounds and generations, subjects in the sample represent some of the renowned figures in modern science. The purpose of studying these women is to understand the factors and forces behind their climb to the top of the scientific establishment —to discover what factors allowed them not only to achieve, but also to be recognized for their contributions during periods when women were not expected to have careers.

Because so many potential variables are involved in career development and advancement, we would be better off focusing on a few key aspects. I considered factors that affect the career development and advancement of female scientific pioneers. Each chapter covers a different aspect of the production of female scientific pioneers: origins, development, and advancement. Using discussions in this chapter as a guide, I looked for patterns of similarities and differences across these ten cases. In the end, I attempted a theoretical synthesis of individual, structural, and institutional explanations.

It is imperative to find out how and why some women are capable of remarkable achievements despite the wide range of personal and professional obstacles that confront them. If what is discovered is applied, a rich, creative, and previously underutilized source of scientific knowledge can be tapped.

## Chapter Three

# Origins: Individual Attributes

This is the first of three chapters looking at different aspects of the production of science makers. Each chapter is devoted to assessing a particular explanation of the science gender gap. To find out why the ten notable women profiled in these chapters succeeded in this male endeavor, I investigated possible reasons for their remarkable successes in spite of repeated failures. I also propose an explanation for this, focusing on a science career as an outcome of individual attributes, structural opportunities, and institutional support.

Drawing from existing ethnographical and biographical data, I used a life course approach to analyze the life and career histories of these women, and integrated the data with the general theme of each chapter. In addition, I incorporated quotes from published accounts of these exceptional women and significant others into discussions to give flesh and meaning to the analysis. The makeup of these exceptional female scientists provides support for the view that the production of female scientific pioneers is a continuous, dynamic process of choice, design, and adaptation.

In this chapter, we try to obtain a substantive understanding of the origin of female scientific pioneers from the nurture argument. To what extent does the sample in this study share the characteristics of pioneers in terms of individual attributes and background? An examination of their life and work histories revealed that they fit the assumptions and predictions of the individual choice approach. Rather than being a passive agent, the ten women in this study actively made conscious decisions on educational and career progress in light of their own interest and circumstances.

Additionally, the results of this analysis take much of the mystery out of the production of female scientific pioneers. The factors and circumstances leading to the career success of men in science might be different from those for women. Due to differences in historical experience, the educational and career opportunities that are available to men and women are invariably different. The general perception is that, by doing what many said could not be done, exceptional female scientists are deviants from societal norms (i.e., "superwomen"). Contrary to expectation, however, they are not deviants from the norms in the scientific establishment. As shown in the discussions below, the ten women in this study share many of the characteristics of pioneers chronicled in Chapter 1.

## HOW TO BECOME MASTERS OF SCIENCE

According to the definition in *Webster's Third New International Dictionary* (1981), science is "accumulated and accepted knowledge that has been systematized and formulated with reference to the discovery of general truths or the operation of natural laws." Simply put, the scientist's mission is discovery of knowledge. Throughout their lives, subjects in this study were preoccupied with the production of new ideas or innovations. Results of their pioneering work have been used by others to understand and solve problems.

Several recurring themes emerge from the life and work histories of subjects in this study. First, why did these women study science? They wanted to study science simply for the pleasure of uncovering the mysteries of the universe. They also showed an early interest in exploring the nature of the world and its inhabitants. Researching on topics or going to places where no one had ever visited presented a challenge to them, yet also constituted an opportunity for them to generate useful scientific knowledge to better humankind.

Second, what did these women find rewarding about the work of science? Their passion for science was derived from a desire to solve the "puzzle." Doing science in and of itself became the intrinsic reward for satisfying their natural curiosity. Unlike their counterparts in other professions, the previously mentioned "gold quest" and "blue ribbon" were definitely not the reasons behind their great accomplishments.

Third, what was the relationship between these female scientific pioneers and their work? As shown in this chapter, life was work and work was life for these extraordinary women. In scientific work, they were able to exercise their intellect and stretch their imagination, and most important of all, had the opportunity to contribute to scientific knowledge.

Not everyone who engages in scientific endeavor will ultimately be recognized as a scientific pioneer. How is it possible to succeed in science in the twentieth century? To help answer this question, this chapter highlights the similarities and differences in background and characteristics of ten female scientific pioneers. Yet the prevailing theme is: *Male and female scientific pi-oneers may have similar traits underlying their success, yet they have to over-come numerous (yet different) barriers at different stages of their careers.*

What are the attributes that make one person a scientific pioneer and an-other person not a scientific pioneer? What are some of the qualities that en-able a woman to get support and resources out of the society, sponsors, and mentors? In addition to having a passion for science, female scientific pio-neers share many attributes considered essential in doing pioneer work (Har-gens 1978; Michalko 1998; Moulin 1955; Ochse 1990; Simon 1974): Scien-tific pioneers are usually high achievers who have succeeded in overcoming many obstacles in family, school, and work. But again, results of these stud-ies are primarily based on the experiences of men.

The notable female scientists in this study also share many critical ele-ments of success. Very early in their lives and careers, their actions and behavior reflected intelligence, optimism, courage, hard work, discipline, in-dependence, enthusiasm, drive, competitiveness, ability to focus and concen-trate, ambition, dedication, confidence, assertiveness, creative thinking and work styles, risk-taking, and perseverance. Many of them showed exceptional academic abilities in childhood. Being a woman in a man's world, all had qualities that helped them overcome obstacles in life and work. Even when they faced turbulent or stressful times, all of them were persistent and con-sistent in how they expressed their purposes and values of life. Constantly hungering for more knowledge, female scientific pioneers always took the initiative to improve themselves through training and development. Their work habits and level of energy surpassed those of their male and female counterparts. Regardless of family background, they all displayed the ability to adapt to changes and deal with adversity and setbacks. Even if they did not receive what they expected or deserved, they did not become discouraged or cynical. Willing to take risk and face loss, they were prepared to make sig-nificant personal and professional sacrifices in pursuit of science.

Not all women who share the above characteristics can be scientific pio-neers. However, female scientific pioneers tend to have these traits described above, and in the following studies:

*1. Marie Curie (1867–1934): Physicist and radiochemist;*
*Nobel laureate in physics (1903) and chemistry (1911)*

Marie Curie's optimistic outlook on life shed light on her approach to prob-lems in life and work: one "cannot hope to build a better world without im-proving the individuals. To that end each of us must work for his own im-provement and at the same time share a general responsibility for all humanity" (Curie 1936:168). Curie tended to view adversity and obstacles in life as challenges and opportunities for change, which is typified by her following statement:

> Life is not easy for any of us. But what of that? We must have perseverance and above all confidence in ourselves. We must believe that we are gifted for something, and that this thing, at whatever cost, must be attained (Curie 1937:121).[2]

Curie acted on her own convictions. While working as a governess in Warsaw, she continuously prepared herself for higher education, however remote the chances of success. Of this effort, she stated, "When I find myself quite unable to read with profit, I work problems of algebra or trigonometry, which allow no lapses of attention and get me back into the right road" (Curie 1937:73). When Curie was attending a private school in Russian-controlled Poland, she was required to suppress her nationalistic views and was called upon to discuss Russian culture in Russian with an inspector. Despite studying in an intimidating environment, she found her high school training positive: "In spite of everything, I like school . . . and even that I love it" (Curie 1937:35). After completing their high school education, Curie and her older sister Bronya wanted to continue their education. Even though both of them had excellent academic records, the Russian government prohibited women from attending a university. Instead of abandoning any hope of acquiring higher education, Curie came up with the idea that she would take a governess position to support her sister Bronya through medical school in Paris. Her sister in turn would help support Curie's university education later on. Curie worked four years as a governess with three different families.

Marie Curie chose radioactivity as the subject of her doctoral dissertation as well as the focus of her future research. Henri Becquerel had already discovered rays released from uranium. However, his discovery had drawn little attention from the scientific community. Curie seized the opportunity to conduct research on this neglected topic. In addition to curiosity, she decided to explore a new phenomenon because she did not have to read a long bibliography of scientific articles to prepare for her work. She could start performing experiments immediately (Pasachoff 1996:35–36). To facilitate her search for rays from other elements, she used a sensitive gold-leaf electroscope rather than the slow and less responsive method of the darkening of photographic film (Rayner-Canham and Rayner-Canham 1998:99). Her risk-taking behavior paid off handsomely. First, her dissertation-examining committee declared *Researches on Radioactive Substances* the greatest scientific contribution ever made by a doctoral dissertation (McGrayne 1998:25). Second, the impact of her choice of subject of study as well as her distinctive approach to research was profound, both for her career and for scientific research. Her discovery of radioactivity brought her fame. It also broke new ground for research in the field of physics.

Curie achieved all this despite numerous constraints and setbacks. For example, Marie Curie and her husband, Pierre Curie, conducted research in a shed which was described by the great German chemist Wilhelm Ostwald as "a cross between a stable and a potato cellar" (Reid 1974:95). Based on her estimation, a properly equipped lab would have allowed them to compress four years of work into one, and therefore minimize their exposure to radiation.

In addition to being a widow, single parent, and independent female scientist for 28 years, Curie faced problems shared by other aspiring female scientists, such as being excluded from the most prestigious scientific institution in France—the Academy of the Sciences. Yet her tenacity, perseverance, and passion for science, among other attributes, were noted by her daughter and others. Irene Joliot-Curie once remarked, "It was my mother who had no fear of throwing herself, without personnel, without money, without supplies, with a warehouse for a lab, into the daunting task of treating kilos of pitchblende in order to concentrate and isolate radium" (Quinn 1995:154). According to Abraham Pais, "Marie Curie was a driven driven [sic] and probably obsessive personality, who should be remembered as the principal initiator of radiochemistry" (Simmons 1996:130–131). Santiago Ramon y Cajal (1999:37) observed:

> That eminent woman, Madam Curie, provides an eloquent example of untiring perseverance. After discovering the radioactivity of thorium, she was unpleasantly surprised to learn that the same observation had been announced a short time earlier by Schmidt in the *Wiedermann Annalen*. Far from disheartened, however, she continued her research uninterrupted . . . she undertook . . . a series of ingenious, patient, and heroic experiments that were rewarded with the discovery of a new element, the remarkable radium. Its properties inspired a great deal of further work that has revolutionized chemistry and physics.

All this suggests that Marie Curie had all the right qualities to initiate and pursue pioneering work. Her responses to setbacks were typical of those who survived and succeeded in science (e.g., Albert Einstein).

## 2. Irene Joliot-Curie (1897–1956): Radiochemist; Nobel laureate in chemistry (1935)

Irene Joliot-Curie is a classic example of the female version of "like father like son." Being brought up in a scientific family, Joliot-Curie showed scientific talents early on, and set her sights for what was important; she concentrated and focused on what interested or pleased her—science. Her life and career histories resemble that of her mother, Marie Curie, in many ways.

Joliot-Curie: (a) studied mathematics and physics, (b) worked in the same field (radioactivity), (c) chose alpha rays of polonium (an element discovered by her mother) as her doctoral topic, (d) married and collaborated with a scientific spouse and had children, (e) worked at the Radium Institute founded by her mother during her entire career, (f) was rejected for admission into the French Academy of Sciences, (g) was appointed professor at the Sorbonne in 1937, and (h) received a Nobel Prize with her husband.

Irene Joliot-Curie applied several times, unsuccessfully, for admission to the French Academy of Sciences. When asked about these repeated failed attempts to gain membership into the most prestigious scientific establishment in France, she took her failure in stride, saying, "Well, at least they are consistent in their thinking!" (Crossfield 1997:122). Her "expect failure but keep trying" approach may well have been a deliberate attempt to underscore the academy's exclusion of women. Joilio-Curie's actions reflected the creative use of an assertive but conservative approach to deal with patriarchy by educated women at the time. This was a tactic combining two political strategies used by American women during the Second World War period: (a) nominating women candidates for offices and complaining later when they were not appointed and (b) confronting or directly challenging government officials and business representatives (Rossiter 1995:22). Another example of the effective use of personal stoicism is that, nearly all her adult life, Joliot-Curie had to cope with a life-threatening case of tuberculosis while she was a leading scientist, wife, and mother.

Working with Joliot-Curie, her prospective husband, Frederic Joliot, saw Pierre Curie in her: "I found in his daughter the same simplicity, common-sense, and ease, who in many ways was the embodiment of what her father had been" (Crossfield 1997:109). Marie Curie had already observed these same qualities and devotion to science in her daughter's childhood, and later at the Radium Institute. Joliot-Curie had the opportunity to demonstrate her scientific skills at a very young age (Crossfield 1997:109; McKown 1961:61). During the First World War, as a teenager, she taught veteran military officers how to use x-rays and geometry to save lives of soldiers. Marie Curie showed confidence in her daughter's independence by leaving her alone working with the military at the war front (McGrayne 1998:117). Joliot-Curie's ability to handle polonium was considered superior to her colleagues at the Radium Institute. She helped train a future Nobel laureate (and her husband) Fred Joliot when he was under her supervision at the institute, and completed a number of original research projects after working six years at the institute.

Despite her privileged childhood, Joliot-Curie's life was not smooth sailing. After losing her father at the age of nine, she had to cope with the stress

of her widowed mother's grief and, subsequently, the public's criticism of her
mother's relationship with fellow scientist Paul Langevin. But none of this
diminished her devotion to science. Joliot-Curie was drawn into the Radium
Institute every day, from morning until day's end.

Like her mother, Irene Joliot-Curie exemplified many of the traits of pioneers. On the other hand, a careful examination of her career trajectory suggests that hers might have been a case of downward mobility in science. One
can make the argument that Joliot-Curie started out great, but shrank over
time and let her husband eclipse her (Bensaude-Vincent 1996).

### 3. Margaret Mead (1901–1978): Cultural anthropologist; President of American Anthropological Association (1960); President of American Association for the Advancement of Science (1975)

Margaret Mead was considered by many of her peers and friends as a bright,
energetic, and daring woman (Gardner and Laskin 1995:70). In light of the
amount of her publications, speeches, and appearances for professional and
general audiences, it is clear that her energy level and work habits were second to none. At the age of 24, she declared:

> I'll be able to come and go and do the kind of work I most want to do. I'm go-
> ing to continue working here [the American Museum of Natural History] as long
> as I can walk up the steps to this tower (Grinager 1999:17).

Mead joined the museum as an assistant curator in 1926, rose to the rank
of curator in 1964, and remained a curator emerita from 1969 until her death
in 1978. Despite being diagnosed with pancreatic cancer in her midseventies,
she carried on her busy work schedule as well as research, and started even
more projects. A few days before she turned 71, when asked if she planned to
slow down then, she replied:

> I won't stop. Have you read about the research on the aged Russians? The ones
> who lived to be over a hundred? They lived longer because they had work to do
> and everybody expected them to do it. So they kept going and stayed alive. It's
> the same with me. I'm working as hard as ever (Grinager 1999:126).

I Made Kaler, Mead's secretary in Bali, attributed his lifelong hard work to
the inspiration of Mead's amazing energy and long hours of work (Mark
1999:68). Not surprisingly, Marvin Harris, an anthropologist, regarded Mead
as "the busiest, hardest-working incarnation of the Protestant ethic since
Calvin" (Sulloway 1996:161). Colleague Ted Schwartz once called for a
"Manhattan project to study the source of [Mead's] energy, her creativity, and
her appetite for and ability to encompass the complexity of very many lives

within her life and intellect" (Gardner and Laskin 1995:83). When asked from where she drew her stamina, Mead replied, "I've had an abundance of energy all my life. My mother was the same way" (Grinager 1999:150).

Mead stood out among anthropologists in several additional aspects: (a) choice of topics for study and research, (b) application of fieldwork methods to make careful observations, (c) blending of different approaches in her work with the use of new techniques, and (d) research productivity. Her excellent communication and diplomatic skills also worked to her advantage. Further, like Rachel Carson, Mead had a distinct writing style that allowed her work to reach vast audiences. Her first book, *Coming of Age in Samoa* (1928), which speaks to both professionals and laypersons, made Mead a household name in America at the age of 30. Mead's ability to popularize ideas and bring people together is best summarized by the following account:

> [She had] the leader's gift for speaking to a collection of diverse persons, many of whom would have disagreed with one another on specific issues, and yet leaving almost every person with the feeling that she had honored his or her perspective. And she had a soothing way of expressing novel ideas that made them congenial, rather than alarming, to those who might be inclined to be hostile (Gardner and Laskin 1995:75).

All this, along with her charisma, contributed to Mead's ascension as a leader in her professional field and a strong social advocate as well.

However, Mead was not without detractors. She lived on a U.S. naval base in Samoa and did not learn the language. So her fieldwork was considered tame and limited by her peers. Her bombast alienated many anthropologists. All this explains why Mead did not win many honors until very late in her career.

In any event, Mead proved herself a risk-taker and determined person when she turned down the proposal of her teacher, Franz Boas, to study an American tribe and instead chose to conduct her first independent fieldwork in the South Seas.[3] Her research was the first of its kind to evaluate the relative importance of biological and cultural determinism of sex-typed behaviors: that if hereditary determines what is possible for males and females, then every society should have similar gender-role arrangements. Determined to punch holes in this prevailing argument, she became the first anthropologist to study women and children in a cross-cultural perspective. Her father, who respected her resolve, paid for her first trip to conduct fieldwork in the South Pacific. At 23, Mead set sail for Samoa to study the life experiences of young women in a remote island not visited by any American anthropologist before. During her entire career, she conducted two dozen field trips of major expeditions.

Mead's collaborative and transdisciplinary approach to studying women and childrearing in remote places was innovative and fruitful. She also brought many new recruits into anthropology. During her entire career, Mead worked with numerous graduate students and young fieldworkers to produce work based on their research on nonwestern cultures.

Mead's scientific marriages[4] also worked to her advantage. First, spousal collaborations allowed her to conduct what she called "complementary and noncompetitive" fieldwork in remote, dangerous places with mutual protection and assistance. Second, such partnerships, which drew on the strengths, insights, and skills of each partner, produced insightful anthropological work on cross-cultural patterns of gender-role socialization. She collaborated successfully with two of her three husbands in New Guinea and Bali, and Mead's collaboration with her third husband, Gregory Bateson, during the 1930s is a case in point: Photographic recording and presentation of cultural anthropological work was the first of its kind at that time. Mead (who excelled in active, extensive note-taking and holistic interpretation of minute details) and Bateson (who focused on passive recording of different sociocultural aspects of primitive people) pioneered the innovative use of ethnography, photography, and videotaping in anthropological fieldwork.

## 4. Barbara McClintock (1902–1992): Geneticist;
## Nobel laureate in physiology and medicine (1983)

Barbara McClintock's life shows that past academic performance is not necessarily a reliable predictor of future educational attainments and career success in science. At Cornell, McClintock found a socially and intellectually stimulating environment, yet she earned just under an overall B average as an undergraduate. However, McClintock's risk-taking and persistence were recognized early on when she chose to focus her research on maize, an out-of-the-mainstream research area.

She became expert at growing and breeding maize for two reasons: not only had she recognized its research significance, but she also preferred to work alone with little help from others. She took care of her corn stocks personally, and spent hours at her microscope. This practice allowed her to see what no one else saw.

Besides being absorbed in the problem at hand, she was meticulous, systematic, and well-organized. Due to her independent and unsophisticated style of research, McClintock stood out among her contemporaries.

One reason for her little need for help from the outside world can be attributed to her brilliance as well as her capacity to pay attention to what was important and ignore what was not.

Marcus Rhoades was intrigued by the ways she conducted research:

> One of the remarkable things about Barbara McClintock's surpassingly beauti-
> ful investigations is that they came solely from her own labors. Without techni-
> cal help of any kind, she has by virtue of her boundless energy, her complete de-
> votion to science, her originality and ingenuity, and her quick and high
> intelligence made a series of significant discoveries unparalleled in the history
> of cytogenetics (Fedoroff 1996:591).

These traits named by Rhoades are also shared among many female scientific pioneers. Marcus Rhoades concurred, stating, "I've known a lot of famous scientists. But the only one I thought really was a genius was McClintock" (Keller 1983:50). Comfort observed, "She was a wizard in being able
to focus on what was significant when there were an awful lot of spurious
things going on. . . . No one else could do the things she could"
(2001:127–128).

One can also make the argument that McClintock made a virtue out of necessity, in that she worked doubly hard in the only way open to her to outdo
the competition. In the 1930s, many men, like Thomas Hunt Morgan at Caltech, had already received large grants and big research teams who went on
to win prizes and honors. The field was on its way to becoming "Big Genetics," which put lone investigators such as McClintock at a structurally big dis-
advantage.

Nevertheless, McClintock's passion for science and commitment to work
was notorious in the scientific community, and her drive and work ethics were
unparalleled in her field. She maintained a high level of enthusiasm for work
during her entire career. To solve a problem, she worked in spurts, night and
day for weeks. She appeared to enjoy what she was doing—challenging the
establishment. She was so good at it that she often found herself solving problems for her peers, teachers, and seniors in her professions. She also kept up
with the literature in and outside her field—a practice she sustained throughout her life. Only very late in life did she change her workday from 12 to 8
hours, and like most female scientific pioneers, she continued her research
until she died.

What kept her and other subjects in this study going was the intrinsic pleasure of doing science as opposed to the tangible rewards. McClintock's
determination and perseverance was evident when she spent six years to
develop an explanation for transposable elements. She also had to overcome
the obstacles experienced by all female scientific pioneers. Thomas Hunt

Morgan, director of the biology division at Caltech, noted, "She is sore at the world because of her conviction that she would have a much freer scientific opportunity if she were a man" (McGrayne 1998:160).

Nevertheless, McClintock held a refreshing attitude about barriers against women in society. For her, gender did not matter over time: "When a person gets to know you well, they forget that you are a woman. . . . The matter of gender drops away" (Keller 1983:76). McClintock's remarks allude to her (or every woman's) ability to transcend gender barriers. Her experience was consistent with the contact hypothesis. Proponents of this hypothesis postulate that interracial contact between people (or between sexes) of equal status in cooperative situations leads to a reduction in prejudice and stereotypes. Conversely, interaction between people of unequal status in competitive circumstances may increase tension and hostility. McClintock's characterization implied women's improved status in the scientific community, and she herself managed to circumvent gender barriers in light of her scientific and professional achievements. In 1944, McClintock became the first woman elected president of the Genetics Society of America, and was named to the prestigious National Academy of Sciences. During the 1950s and 1960s, McClintock turned down teaching and research opportunities elsewhere and remained at Cold Spring Harbor on Long Island, New York. This proved to be a proper decision, given her repeated unpleasant experiences at the University of Missouri and other academic institutions.

McClintock's self-confidence and self-motivation were unparalleled when her novel concept of transposable elements perplexed the scientific community for decades. This gung-ho attitude originated when she learned to "handle the fact of [her] difference in high school." McClintock's own account revealed her distinctive approach to problems in life and career:

> I found that handling it in a way that other people would not appreciate, because it was not the standard conduct, might cause me great pain, but I would take the consequences for the sake of an activity that I knew would give me great pleasure. And I would do that regardless of the pain. . . . It was constant. Whatever the consequences, *I had to go in that direction* [italics in original text] (Keller 1983:27–28).

Her persistence and perseverance paid off. As Comfort observed:

> By quiet persistence and personal integrity, she outlasted the narrow-minded[ness] and bigotry that blinded her colleagues. She seems in every way the opposite of the archetypal molecular biologist: senior, humble, intuitive, and female working in the fields on a large, slow-growing, complex organism—in contrast to the young rational, arrogant, male biologists working on bacteria and viruses (2001:2).

Like her female peers, McClintock developed a lifelong proactive strategy to deal with professional setbacks: "Getting into something else, some new knowledge. . . . It takes me away from brooding about myself" (Keller 1983:86).

McClintock was the opposite of her contemporaries in terms of research orientation, approach to problem solving, and work style. In these attributes, she stood out among her peers and rivals. However, in many respects, her tactics are no different from those adopted by Mead and other female scientific pioneers in this study. Her style was characterized by thoroughness, rigor, integration, and synthesis. According to McClintock, one must "have the time to look, the patience to 'hear what the material has to say to you,' the openness to 'let it come to you,' and have 'a feeling for the organism'" (Keller 1983:197). Equally important, she thought like other male scientific geniuses (Michalko 1998; Simonton 1999).

*5. Maria Goeppert-Mayer (1906–1972): Mathematical physicist; Nobel laureate in physics (1963)*

Similar to Margaret Mead, Maria Goeppert-Mayer avoided direct competition with her male counterparts in physics. Her modest approach to research and publication won her friends and loyal supporters as well as respect from her colleagues. This approach seemed to fit into her role as a female scientist in a male-dominated discipline. Despite her preference for "associating with the boys," Maria did not display the (masculine) trait of competitiveness expected of male and successful scientists. This observation corresponds to the humble image of Goeppert-Mayer held by others. As noted by Jones, "She never developed the charisma we often associated with scientific greatness" (1990:203).

Her humble demeanor is probably one reason why her scientific talents were initially underestimated in the scientific community. When he heard the news of her winning the Nobel Prize in 1963, a physicist who had known the Mayers during the 1940s said, "I always thought Maria was competent. But in those years if you had asked me about her chances of winning a Nobel Prize, my answer would probably have been, 'Are you kidding?'" (Dash 1973:233). When urged by her husband, Joseph Mayer, to publish her paper on the shell model of the nucleus, Goeppert-Mayer was extremely reluctant to submit it for peer review because of her concern for certain complications after seeing preprints of papers by two other physicists. Her response was typical of her style:

I will wait. I will write to the editors of *Physical Review* and ask when those papers will come out, and send them something that I will ask them to print at the same time. I must not take advantage because I saw their preprints (Dash 1973:319).

In keeping with her reluctance to grasp the spotlight, Enrico Fermi had to turn down her invitation as coauthor of her work. Her peers attributed Goeppert-Mayer's modesty to her privileged background. Competition among experimental physicists might have contributed, in part, to her secretiveness or reticence. On the other hand, this style reflected her rigor and thoroughness in her work. She reminded her students that one is bound to discover his errors when he publishes a paper based on false assumptions (Dash 1973:321), and cared more about building a good scientific reputation than rushing for scientific recognition. Her approach to publishing is consistent with the observation that some scientists are reluctant to publish their findings until and unless they are convinced of their validity, a characteristic not unique among female scientists (Hagstrom 1974:10).

Many were impressed by Goeppert-Mayer's scientific accomplishment in light of her modesty and being a late bloomer. For example, Mitchell Wilson, a student of Fermi and a physicist-turned-novelist, said:

> Her achievement was all the more remarkable because she had done her work when she was well into her forties and she had only recently come into the field of physics from chemistry, and most of all because she was a woman (Dash 1973:233).

All this was done in the context of working as an unpaid volunteer professor at various academic institutions for decades before receiving the Nobel Prize. That meant that, unlike her male peers, Goeppert-Mayer did not receive such tangible benefits as salary, proper credits, and recognition for years of teaching, advisement, and college services. More importantly, of the female scientific pioneers in this study, she is the one who worked the longest without pay and formal recognition. Because of the lack of a "real" job for decades, Goeppert-Mayer literally became a "serial volunteer researcher/teacher," a predicament many postdocs face in the science fields today.

If antinepotism rules posed difficulties for women pursuing scientific careers at the time, they did not seem to bother Goeppert-Mayer. Her daughter, Marianne, did not recall her mother complaining about any discriminatory practices. The respect and admiration she received and the opportunity to work with a group of leading scientists at top research institutions excited Goeppert-Mayer and kept her going. At any rate, she was more concerned with balancing work and family than dealing with the restrictions on her career progress. As Fay Ajzenberg-Selove recalled:

> In 1955, the life of a woman physicist was very rough. When she heard that I was getting married, Maria Goeppert-Mayer, who saw me at a meeting. . . . told

me that while it was hard to be a woman physicist, it was nearly impossible to be a married woman physicist (1994:114).

Goeppert-Mayer ignored these predicaments and instead used the time and opportunity to learn new knowledge in other fields, such as experimental chemistry from her husband, Joe Mayer, and Karl Herzfeld, and nuclear physics from Edward Teller and Enrico Fermi.

Goeppert-Mayer's patience and lifelong interest in learning, rather than fighting the system, proved to be bold and strategic moves in light of her winning of the Nobel Prize for the shell model of the nucleus. Her resolve was put to test when she had to learn nuclear physics from scratch and work with renowned figures in the field like Fermi and Teller. When asked why she continuously updated her skills, she reported being motivated by economic and professional reasons (McGrayne 1998:185–186). Her persistence and curiosity paid off when Fermi accidentally posed the question, "What about spin-orbit coupling?"

The knowledge and skills Goeppert-Mayer acquired over the years, plus insights like these, became invaluable to her later work on nuclear structure.

*6. Rachel Carson (1907–1964): Marine biologist; National Audubon Society Medal (1963); American Geographical Society's Cullem Geographical Medal (1963); Presidential Medal of Freedom (1980); Issue of Rachel Carson stamp (1981)*

Rachel Carson also shared many traits of scientific pioneers. She was always curious about the world around her, with a particular interest in nature and wildlife. Solitary isolation did not bother her because, like Curie, Mead, and Yalow, as a child Carson had a love of books. Reading and thinking were her passions. These also gave her the opportunity to develop an analytical mind, a capacity for original insights, and literary talent. Carson's mother and teachers noticed her scientific interests and abilities early on. Her meticulous research, literary talents, and concerns with environmental issues contributed to her development into a serious science writer: "I am that comparatively rare phenomenon, a scientist who is also a writer" (Lear 1998:115).

Her determination set her apart from peers at the Pennsylvania College of Women. Over the objection of the college's administration, she changed her major from English to Biology. At that time, no one, including, Carson could imagine that combining biology and writing was a realistic career option, especially for a woman. She understood and anticipated the challenges ahead of her:

Writing is a lonely occupation at best. . . . During the actual work of creation the writer cuts himself off from all others and confronts his subject alone. He moves in to a realm where he has never gone before—perhaps where no one has ever been. It is a lonely place and even a little frightening. No writer can stand still. He continues to create or he perishes. Each task completed carries its own obligation to go on to something new. I am always more interested in what I am about to do than in what I have already done. . . . In some mysterious way the subject chooses the writer—not the other way around (Lear 1998:287).

Her career success suggests that it was a creative and right choice.

Carson's career success can also be attributed to selecting a very important topic for study—environmental issues. But what prompted her to expose the profound impact of subtle, yet dangerous human actions on ecological life was a sense of social responsibility. Taking a risk that few novice or even established scientists could afford to take, Carson wrote about the indiscriminate use of DDT and other toxic chemicals in pesticides and weed killers.[5] The result of her meticulous research on the impact of toxic substances on the environment was the publication of *Silent Spring* (Carson 1962). This book received a great deal of attention from the media and government, praise from the public and professionals, but also severe criticisms from the chemical industry. Carson's risk-taking, determination, and perseverance were rewarded when the government finally acted to control the use of pesticides.

Carson's statement on her outlook on life and the future was very similar to that of Marie Curie's: "In order to achieve one must dream greatly—one must not be afraid to think large thoughts" (Lear 1998:297). Carson's lifelong concern with environmental issues was a manifestation of these bold dreams; she was not only interested in the effects of pesticides on the environment, but also concerned about what could happen to its inhabitants if the indiscriminate use of pesticides continued (McCay 1993:67).

With a master's degree in zoology from Johns Hopkins University, Carson worked as an editor of wildlife publications for the federal government. This job allowed her to observe nature and do research and writing on the side. However, unlike other female scientific pioneers in this study, Carson had neither the highest academic honors in her field nor the advantages of university affiliation to back her up against attacks and criticisms on her works by the scientific community and industry. This perceived lack of proper scientific credentials led to her meticulous and systematic approach to all research endeavors. For example, in *The Sea Around Us*, Carson (1951) used more than a thousand sources and corresponded with oceanographers and experts all over the world. She also used two or three authorities to support each statement in *Silent Spring* (McCay 1993:80). Her style of continuously seeking advice was common among the female scientific pioneers in this study.

An inclination to listen and confer with experts was vital to the success of her book projects. Carson took the initiative to approach editors. She was also assertive in her financial negotiations with publishers. Because the audience for *Silent Spring* encompassed scientists, chemical industries, and government agencies as well as laypersons, Carson included 54 pages of principal sources in a 368-page book to make her case about the potential impact of arbitrary use of pesticides on wildlife and humans. In addition, sixteen experts read and commented on the manuscript (1962:ix–x).

Carson liked to work independently, but not in isolation. The success of her book projects can be attributed in part to a habit of consulting with a range of experts. Yet despite her frequent consultations with experts, Carson never had doubts about her qualifications as a scientist. For instance, on the dust jacket of *Silent Spring*, she wanted to include the fact of her studying genetics with H.S. Jennings and Raymond Pearl at Johns Hopkins, as well as studying radiation as a cause of mutation (Lear 1998:398).

## 7. Rita Levi-Montalcini (1909– ): Neuroembryologist; Nobel laureate in physiology and medicine (1986)

Besides sharing the qualities of Carson and others, Rita Levi-Montalcini attributed her success in science to a highly developed intuition: "I have . . . just average intelligence. [But intuition is] something that comes to my mind. . . . It is a particular gift, in the subconscious" (McGrayne 1998:213). But in fact, she is *very* smart. For example, she and her cousin Eugenia made the highest score after studying just six months together for the University of Turin entrance examination.

Having an ability to underestimate the difficulties of present tasks is another factor:

> I have become persuaded that, in scientific research, neither the degree of one's intelligence nor the ability to carry out one's tasks with thoroughness and precision are factors essential to personal success and fulfillment. . . . Most important for the attaining of both ends are total dedication and a tendency to underestimate difficulties, which cause one to tackle problems that other more critical and acute persons instead opt to avoid (Reynolds 1999:103).

Conducting experiments in a makeshift lab set up in her tiny bedroom was a testament to her claim:

> I had practically nothing—just eggs and an old microscope. . . . I used a sewing needle I sharpened myself on a stone. I kept things sterile with alcohol. . . . It was a pure miracle that I succeeded with such primitive instrumentation, and it cannot be repeated. . . . Today a scientist could not go far if he had only a bedroom and a microscope. It would be ridiculous (Liversidge 1988:73).

This is a perfect example of how she and other female scientific pioneers in this study could manage to accomplish so much in scientific research with so few resources and little support.

Like Curie, Joliot-Curie, McClintock, and Carson, Levi-Montalcini admitted that she was an introvert and lacked proper social skills:

> A natural aversion to sports, and an even stronger one to establishing contacts with girls of my own age, accentuated my profound sense of isolation, which derived in part also from timidity and my total lack of inclination to approach young men of my age or older in the expectation of meeting a future life companion (Levi-Montalcini 1988:35).

But that did not have any impact on her working with other scientists, such as (a) collaborating with leading neuroembryologist Viktor Hamburger, biochemist Stanley Cohen, and others; (b) working with young postdocs; (c) carrying out joint research that eventually led to her winning the Nobel Prize; and (d) sharing administrative responsibilities of running the labs in Rome and St. Louis, Missouri with her compatriot, Pietro Angeletti.

Levi-Montalcini's drive, determination, and perseverance were put to the test when she encountered racial and gender barriers during the war. First, she and other Jews were banned from doing any academic work or pursuing any employment in universities and academies. She lost her job as a research assistant during the Second World War. Rather than becoming disheartened or giving up, she moved the research from her university lab to a lab in her bedroom. Little did she or anyone know that her clandestine bedroom experimental operations served as a springboard for her later discovery of nerve growth factor (NGF). Second, due to anti-Semitism against Jews in Italy, she was prohibited from legally practicing medicine. Third, because of her Jewish name, her work could not be published in any Italian scientific journals. Fortunately, her work eventually appeared in Belgian and Swiss journals that were also circulated in the United States.

To overcome these and other difficulties, she exercised discipline, and paid attention to what was important to her in the long term and ignored what was not. Like McClintock's colleagues' lukewarm reception to her transposable elements, the scientific community initially had very little interest in Levi-Montalcini's NGF. Her self-reflection underscored her strategy of underestimating the difficulties in life:

> Many years later, I often asked myself how we could have dedicated ourselves with such enthusiasm to solving this small neuroembryological problem while German armies were advancing throughout Europe spreading destruction and death wherever they went, and threatening the very survival of western civiliza-

tion. . . . The answer lies in the desperate and partially unconscious desire of human beings to ignore what is happening in situations where full awareness might lead one to self-destruction (McGrayne 1998:210).

Despite her introversion, Levi-Montalcini was an adventurous and independent person. Since 1946, she spent three decades away from her homeland of Italy, first as a postdoc and later as a professor at Washington University in St. Louis, Missouri.

Like many female scientific pioneers in this study, Levi-Montalcini downplayed the significance of gender on career advancement, despite the fact that she and others faced a huge amount of discrimination against women for their time. Her unique attitude served as a self-defense mechanism against outside forces and, more importantly, a reaffirmation of her self-worth as a scientist:

I do not believe my career has been affected by being a woman. A woman's career may be hampered in two ways: by her feeling of responsibility to her family and by professional discrimination. In my case, I had no family obligations because I am single and did not have that problem to face. As for the second difficulty, I have never felt, either in Italy or in the United States, any professional hostility because of being a woman (Wasserman 2000:42–43).

However, her remarks cannot be construed as antifeminist. Of all the female scientific pioneers in this study, her views toward marriage and family are closest to McClintock's. This is perhaps due to a striking similarity in their desire for autonomy and independence. Levi-Montalcini remarked:

I simply could not live with another person and adjust my own life to somebody's else. . . . [A] man, a husband, or children want a lot from you, and . . . I was not always willing to give anything. My work was important to me. . . and I've been in excellent relationships with many people. But never would I have liked to share my life with any of them . . . [I]f you marry a very intelligent man, but if you are equally brilliant, one might resent the other being more successful. Also, they may be forced to go to different places to work . . . It's just a problem when two persons want success or want to devote themselves to work. One has to sacrifice to the wishes of the other, to come second in the marriage. If he is a man, he doesn't like it. Women used to be good at accepting position number two (Liversidge 1988:102).

She, along with McClintock, avoided the potential conflict between work and family by remaining single and not having parenting responsibilities.

Despite her claim of being indifferent to gender bias, Levi-Montalcini was sensitive to the treatment she received from the scientific establishment. For example, she did not like a man named James D. Watson because of his way of treating her:

> Viktor Hamburger at Washington University wanted me to meet the most bril-
> liant of his students, James Watson . . . Our brief encounters over the following
> years were characterized by his absolute indifference toward me—an attitude I
> saw as part of his well-known antifeminism. [However,] I was never bothered
> by it" (Levi-Montalcini 1988:138).

Rita Levi-Montalcini shared the same attitude toward work as Mead and McClintock. Her work was part of her life and vice versa:

> The moment you stop working, you are dead. My brain has not deteriorated
> . . . so why should I not use my brain as I use my hand . . .? My eyes are also
> good, I see very well. As long as I have this advantage, why shouldn't I
> work? . . . For me it would be unhappiness beyond anything else; I couldn't af-
> ford to go on without my work. I don't work for the sake of mankind; I work for
> my own sake (Liversidge 1988:104).

And she believed that if one keeps working at a problem diligently, a solution will finally emerge. Like many other female scientific pioneers, she counseled patience, hard work, and persistence in the following statement: "[We] often neglected [the] rule that many apparently unsolvable problems at one point or another unexpectedly find their solution" (Levi-Montalcini 1988:166). Here, and again, she demonstrated the creative thinking style as identified by Michalko (1998).

## 8. Dorothy Hodgkin (1910–1994): Crystallographer; Nobel laureate in chemistry (1964)

Dorothy Hodgkin was known in her family and among her peers to have intense engagement with everything around her. Like Rita Levi-Montalcini and other female scientific pioneers in this study, she was a master of making minute observations as well as synthesizing divergent information. For instance, Hodgkin's success in making significant discoveries in penicillin, Vitamin $B_{12}$, and insulin rested in her ability to pay attention to what was important and ignoring what was not:

> [There is] a great advantage to start early, [because] one gets a certain amount
> of notoriety from being the first person to do things which anybody else really
> could have done. What I find difficult to know is why more people didn't take
> up this particular method of attacking problems at the same stage as we did
> (Wolpert and Richards 1988:72).

Hodgkin chose to work on interesting and medically important problems. For instance, she was the one to recognize the significance of crystals when browsing through Desmond Bernal's reprints while working in his lab. As noted earlier, being able to focus and concentrate on each task is a common trait among female scientific pioneers. This distinct quality of hers was highlighted by J. Dunitz:

> Dorothy had an unerring instinct for sensing the most significant structural problems in this field. . . . She had the audacity to attack these problems when they seemed well-nigh insoluble. . . . She had the perseverance to struggle onward where others would have given up and she had the skill and imagination to solve these problems once the pieces of the puzzle began to take shape (Julian 1990:376).

Due to extended separation from her parents in childhood and adolescence, Hodgkin acquired the ability to deal with personal problems and professional setbacks independently and confidently. For instance, she was able to catch up with her peers despite trailing academically in high school:

> When I first went to secondary school, I was rather behind. I was terribly behind in arithmetic, and it was only at the end of my time there, at the very end of the last year, that I was first in the form (Wolpert and Richards 1988:76).

These were not the only challenges Hodgkin faced and overcame. Her hands were crippled by arthritis in childhood. However, she did not let the constant pain and discomfort affect her life and work. As a woman, she could not join the chemistry club or attend its small weekly talks about current research at Oxford, which had a reputation as "a man's university" for half a century. Further, because she initially held no official appointment in chemistry or mineralogy, she could not participate in decision making about the future of crystallography at Oxford. Fortunately, holiday visits to Desmond Bernal's lab gave her scientific companionship. Thus, in spite of being excluded from the *inner circle*, she managed to stay on top of the latest developments in her field.

Like Curie, Goeppert-Mayer, and Levi-Montalcini, Hodgkin preferred to tackle her problems with tact and achieved her goals quietly, with little fanfare. First, she denied that being a woman had any adverse impact on her career advancement. Hodgkin used her "lone girl" status at Oxford to secure support for research equipment from male professors. When asked if she felt strongly about the position of women in science, she said:

> No, I think it is because I didn't really notice it very much, that I was a woman amongst so many men and the other thing is . . . that I am a little conscious that

there were moments when it was to my advantage . . . and at the time just after
the war, when there was an air of liberalism abroad and the first elections of
women to the Royal Society that were made, that probably got me in earlier than
one might have as a man, just because one was a woman (Wolpert and Richards
1988:77).

Second, Hodgkin did not let marriage and family responsibilities interfere
with her work: "To speak more personally, I could never have carried out the
amount of scientific research I have achieved if I had not, at the time of my
marriage, been earning a sufficient salary to permit me to pay for help in our
home" (Ferry 1998:248). However, because of the lack of stable employment
of her husband, Thomas Hodgkin, Dorothy Hodgkin was the primary bread-
winner for her family. As a result, like many of her male counterparts at the
time, she had the double duty of supporting a family while pursuing a scientific career.

Third, Hodgkin had developed a reputation as a "good" lab worker—
dependable, skillful, and diligent. Like Maria Goeppert-Mayer, she would not
submit a research paper for consideration until and unless she was certain of
the research findings. Her meticulous approach to problem solving is common among scientific geniuses. She admitted that she liked to "think with her
hands" (i.e., visually).

Finally, Hodgkin's devotion to work and pioneering the technical skills in
crystallography reflected her determination and career ambition:

Nobel Prizes were a bit out of my knowledge. As a young girl I did not know a
thing about Nobel Prizes. I wasn't particularly ambitious. I just liked working in
this particular kind of field. I didn't imagine myself making enormous discoveries (Wolpert and Richards 1988:78).

In spite of her admitted lack of ambition for it, she was nominated several
times for this award and eventually became the sole winner of the Nobel Prize
for chemistry in 1964: the first British female recipient of the award.

*9. Rosalyn Yalow (1921– ): Medical physicist;*
*Nobel laureate in physiology and medicine (1977)*

Rosalyn Yalow was a genius in the eyes of her junior high school teacher. At
home, Yalow was encouraged to follow her interests and inclination. As a result, she grew up as a determined, independent, and assertive woman, which
prepared her to survive and thrive in a man's domain academically and professionally. Her role models were Edith Quimby and Marie Curie. Inspired by
reading Eve Curie's biography of her mother, *Madame Curie*, Yalow wanted
to follow in Curie's footsteps as a wife, mother, and successful scientist

(Straus 1998:103–105). Though Yalow was the younger of the two children in her family, she was considered more masculine than her older brother, Allie. Her assertiveness and combativeness made it possible for Yalow to enter and succeed in the field of physics:

> I, as a small child, made up my mind that people and institutions were going to need me, and I was going to let them know that they would need me. . . . You go where the power is. You don't isolate yourself. You don't need to be protected. It's part of my aggressive approach to things (McGrayne 1998:355; Straus 1998:56).

Others attributed Yalow's success in science to her toughness and assertiveness as well. Her teacher, Maurice Goldhaber, professor of physics at Brookhaven National Laboratory, recalled:

> [Yalow] was very clever . . . aggressive, always in a hurry. In retrospect, since she was so successful, one can say her aggressiveness was built on something, it wasn't just for the sake of aggressiveness. It was because she was going somewhere (Straus 1998:37–38).

Others have made similar observations of Yalow's serious approach to solving problems in life and work. For instance, Sherman Lawrence, her first boyfriend and later long-time family friend, observed that after her father died:

> She took over. . . . it was just like she was in science. She was that way with everything, she made all the decisions. She was meticulous. She was determined. Nothing would stand in her way. She used to scare me (Straus 1998:42).

Lawrence's assessment of Yalow's ability to handle challenges was extremely accurate in light of her predicament as a Ph.D. candidate at the University of Illinois:

> I was the first woman to have a graduate assistantship in physics there since 1917. And I remember when I first got there I was given a list of housing on campus that wouldn't take Jews (Straus 1998:34).

Yalow's engineering background stood her in good stead when she started working part-time as a consultant for the Bronx Veterans Administration (VA) Hospital: "My experience as an electrical engineer was quite useful, in as much as commercial equipment was not readily available; thus much of our early equipment was made by me or according to my design" (Reynolds 1999:130–131).

Those who expected Yalow to directly challenge the scientific establishment would be fascinated and perhaps frustrated by her view toward gender discrimination:

> Personally, I have not been terribly bothered by it. I have understood that it exists, and it's just one other thing that you have to take into account what you're doing. . . . If I wasn't going to do it one way, I would manage to do it another way (McGrayne 1998:338).

She acknowledged its existence, yet did not think that it had *any* impact on her life and work:

> I didn't pay any price for working so hard, and I never felt any gender bias. . . . I'd like to think that I serve as a symbol of the fact that women can make it in what was once a man's world (Straus 1998:256).

In reality, Yalow worked doubly hard in order to achieve parity with men. Instead of proposing structural changes, her strategy was individual: "Women, even now, must exert more effort than men do for the same degree of success [even though] there was something wrong with the discriminators, not something wrong with me" (McGrayne 1998:341–342). This attitude was likely why she hung a sign on the wall of her office at the VA Hospital with the motto: "Whatever women do they must do twice as well as men to be thought half as good. Luckily this is not difficult" (Straus 1998:112). Her combative attitude could not change the fact that as a woman with a doctorate in physics, Yalow could not possibly be a part of the medical center's inner circle: Due to a lack of money, she was unable to pursue medical studies.

Self-reliance was the basic lesson Rosalyn Yalow learned growing up in a poor and uneducated family:

> I wasn't handed college or graduate school or anything else on a silver platter. I had to work very hard, but I did it because I wanted to. That's the key to happiness. . . . A child must learn from the cradle that upward mobility depends on what people themselves do (Reynolds 1999:131–132).

For example, when she wanted braces, Yalow earned money by helping her mother turn collars for a neckwear factory at home. Her attitude was typical among scientific pioneers—facing challenges head-on: "Who had the money? If you wanted something, you worked for it. It didn't keep me from doing my homework" (McGrayne 1998:336).

Another positive trait and accomplishment of Yalow's is her ability and willingness to mentor both male and female scientists. When men were typically the role models and sponsors in the scientific establishment, Yalow became the pioneer in "scientific motherhood" (Reskin 1978:14). Few female scientific pioneers have had the opportunity to and interest in mentoring the next generation of scientists. Yalow is a lone star in her field because of the guidance and support she offered to aspiring scientists over the years: the "professional mother" of many successful male and female scientists. No other female scientific pioneer in this study has produced scores of "professional children" as Yalow has. She managed to mentor many protégés while doing research, running a lab, and keeping a home at the same time. Simply put, Yalow set a precedent in the history of women in science.

Yalow's intense focus and single-mindedness became her trademark in science. Most important of all, like many female scientific pioneers, she enjoyed what she was doing immensely: "What am I going to do? Ride a bike? Play tennis? [The lab] is where the excitement is" (Reynolds 1999:131). When her work with Solomon Berson was rejected for publication by a prestigious scientific journal, rather than complaining and arguing, they chose to focus on what was more important and exciting. Every time Yalow was passed over for the Nobel Prize, her husband, Aaron Yalow, reported, "Her reaction was just 'What do I have to do to win?'" (McGrayne 1998:351). Yalow was keenly aware of the fact that women in scientific partnerships tend to be ignored and their contributions marginalized. That is why she worked hard to claim the credit she deserved, and expected to be treated as men's equal: "When Sol [Berson] died the question was, would a woman Ph.D. who was a partner in a team win a Nobel Prize? Had I died there would have been no question that he would have gotten the prize" (Straus 1998:227). At the time of Sol Berson's death, Yalow was in a situation similar to that of Marie Curie's when scientific spouse/partner Pierre Curie died suddenly. Like Curie, she continued the research and went on to receive the Nobel Prize in medicine and physiology in 1977, as well as numerous honorary and scientific awards.

Yalow did not devote herself to work to the exclusion of other aspects of life. Her views toward work, marriage, and family are the direct opposites of those of Levi-Montalcini, and more in line with those of Marie Curie, Irene Joliot-Curie, Margaret Mead, Maria Goeppert-Mayer, and Dorothy Hodgkin. To Yalow, family and work are not incompatible for women. Like Mead, Yalow believes that (a) female scientists should marry and have children, (b) mothers who have career aspirations should not stay home to take care of children, and (c) the society should provide quality childcare to working mothers (Reynolds 1999:133). When asked if she enjoyed being a grandmother, her response espouses the image of an ideal female scientist:

I think it's important for bright women to get married and have children. I think it's a serious mistake for bright women not to get married. . . . The three women

Nobel laureates after me are not married, and I think that they're missing something. For the most part, men who become Nobel laureates are married and have children. I think that women who become laureates should also be married and have children. After all, Marie Curie's daughter became a laureate" (Straus 1998:110–111).

Aside from being a role model for young scientists, Yalow is a role model for married female scientists, and does not see any potential conflict between work and family:

You can have it all. I can't think of anything in the world that I want that I haven't had. I have my marriage, two wonderful children. I have a lab that is an absolute joy. I have energy, I have health. As long as there is anything to be done, I am never tired (Reynolds 1999:134).

Similar to other scientific pioneers, there is no demarcation line between life and work for Yalow. At 77, she still worked at the VA Hospital twice a week. When asked if she would have done anything differently in her life, she replied, "Nothing" (Reynolds 1999:134).

## 10. Fay Ajzenberg-Selove (1926– ): Nuclear physicist; American Physical Society's Nicholson Medal for Humanitarian Service (1999)

A specialist in nuclear spectroscopy and one of the most cited contemporary physicists, Fay Ajzenberg-Selove has also demonstrated that one could make it in science through determination and hard work: "The signal I was getting from my grades clearly warned me that I was not a born scientist. But I had fun, and I was determined to keep trying" (Ajzenberg-Selove 1994:50). She indeed kept on trying after failing her first-year physics courses at Columbia University. Her stubbornness paid off.

Ajzenberg-Selove's drive and perseverance came from her wartime experience in adolescence (1926–40). Her father showed her and other siblings how to commit suicide with a dagger (with dignity) instead of being killed by the Germans (should that circumstance appear). When recalling the episode, she related, "We were all extremely depressed. . . . As for me, I just wanted to survive" (Ajzenberg-Selove 1994:27). This traumatic episode might have strengthened her resolve toward dealing with both hardships in life and gender bias in employment.

Ajzenberg-Selove came to the United States as a refugee, at the age of 16 in 1940. Like Yalow, she does not shy away from competition or confrontation. For the most part, her no-nonsense and combative attitude got her where she is today. She took lawyers to court to fight for her father's estate. She refused to sign an employment contract when the salary was 15% less than the original offer. When ladies rooms were not available at her new workplace, one was promptly built for her even after she claimed that she would not mind sharing the men's room. She sued her employer and won her tenure case. At 44, she underwent a cancer operation and survived.

While most female scientific pioneers in this study downplayed the significance of gender on career progress, Ajzenberg-Selove's experiences underscore the problems she and other women faced in colleges, professional societies, and scientific establishments. She was well aware of the problems that a female physicist faced in 1955, as revealed in her brief encounter with Maria in an earlier section on Goeppert-Mayer of this chapter. In addition, she stated: "Gender discrimination is a matter of still greater importance in graduate school when relationships with older scientists, with future patrons, are first established—relationships that are critical to a student's entire career" (Ajzenberg-Selove 1994:221).

Fay Ajzenberg-Selove reported working 90 hours a week at Haverford, devoting one-third of her time each to teaching, research and scholarship, and committee work. Like other female scientific pioneers in this study, she gave her best in whatever she did. She declared, "[I] consciously resolved to live a life that I would not regret as I lay dying" (Ajzenberg-Selove 1994:17).

## IMPLICATIONS

Female scientific pioneers come with different sets of personality traits. What is common among them are a set of unique qualities they shared with other scientific pioneers. To become a scientific pioneer, having a passion for science is not enough. One must have more than talent to succeed in science. Among others, a creative style of thinking, diversity in problem selection and problem solution methods, and personal qualities are essential for the successful pursuit of pioneering work.

So, what is the recurring theme in the early and later lives of these notable scientists? Science is a game for anyone willing to subscribe to the norms and practices of the scientific community. In the case of women, stamina to withstand subtle and not-so-subtle barriers is essential. The women profiled here defied cultural and social norms regarding men and women. They seemed to have a positive attitude toward personal problems and professional setbacks: "It is important not to fear from failure, but to learn from it" (Gibbons, Holden, and Kaiser 1996:8). Additionally, irrespective of their social origins and cultural background, all of them had to deal with cumulative disadvantages in their educations and careers. These ten female scientific pioneers all had the ability to render random or uncontrollable factors insignificant. For them, an obstacle served as a propeller or springboard to their success, rather than a deterrent. Whether singly or through collaboration, they knew what they wanted and how to get it.

Equally important, all of them were aware of gender barriers in society and in the field of science. Fortunately, they were independent women who could take care of themselves and operate in different capacities in the scientific establishment. Contrary to expectation, they were not out to prove themselves by consciously breaking barriers for women in science. A relevant issue is that, given the amount of effort, energy, and discipline these exceptional women had put into their study and/or work, wouldn't they have gotten ahead as fast as their male counterparts under similar circumstances? To answer this question, we now turn to the second phase of the production of female scientific pioneers.

*Chapter Four*

# Development: Structural Opportunities

## BEYOND TALENTS

This chapter explores the conditions that make the production of female scientific pioneers possible: historical factors, positive and negative, that can shape one's aspirations for scientific development and a science career.

Many women who possess the important qualities discussed in preceding chapters do not become scientific pioneers. Being outstanding in ability, effort, and style alone is not sufficient to propel a talented woman to the forefront of science. So why are men and women with similar traits and qualities differentially rewarded for their contributions to science? A set of structural forces must be in place to open up new possibilities for women. As Rosenbaum put it, "Fundamentally, structure dominates the career process. . . . [Structural factors are] to be central in mediating the effects of most individual factors on earnings and career attainments and which determine the responsiveness of career trajectories" (1984:263). In other words, selection of individuals for advancement is usually beyond any one person's control; career prospects of both men and women in science can be affected by changes in the society. This assertion is congruent with Helson's conclusion based on the results of her study of creative male and female mathematicians: "The striking differences between creative men and women in professional status and in productivity after graduate school seem to reflect social roles and institutional arrangements more than fundamental creative traits" (1971:219).

Based on her observations of a group of British scientists and engineers working in two large industrial organizations, Evetts draws a similar conclusion. Career trajectories are the products of culture and society. Specifically, cultural and social contexts heavily influence one's career choice and career experiences (Evetts 1996:147). Evetts also notes that organizational restructuring has not only affected career opportunities, but it has also differentially affected the career opportunities for men and women.

These assertions about the production process of female scientific elites are also based upon the observations of women in European science, made by Stolte-Heiskanen (1983). She notes that a series of sociocultural factors have impinged on the status of women in science in Austria, Belgium, Finland, Hungary, Poland, and Sweden. These factors include: (a) the possibilities for women to obtain higher education, (b) employment opportunities in research and other science-related activities, and (c) the norms regarding the status of women in society.

Forthcoming discussions underscore the historical reciprocal relationship of women's prospects in science and the society. Economic and political upheavals in the midnineteenth and midtwentieth centuries provided newcomers (including women) with opportunities to pursue pioneering work in science. The structural shifts that precipitated women's entry into the scientific establishment can be classified into three categories: cultural, economic, and political. An overview is given to illustrate the complex interaction between these collective transformations. To bolster these claims, I will cite specific examples from the life and work histories of the sample.

## THE CULTURAL FACTOR

A series of major cultural developments have facilitated women's formal participation in science. A clear sign of diminishing social constraints on women was the first women's movement in Europe and North America (Rossiter 2003). After the opening of major universities and graduate schools to women in the late nineteenth and early twentieth centuries, women enjoyed new educational opportunities. Needless to say, the situation for women in science varies widely across cultures and time periods.[6] For example, in the United States, Title IX of the Education Amendments of 1972 prohibits sex discrimination in federally assisted education programs. This legislation had the single most dramatic impact on the presence of women in the sciences (U.S. Department of Education 1997).

Gaining access to Ph.D. programs has made it possible for women to pursue serious scientific work. As shown in Table 2, female scientific pioneers in this study were beneficiaries of this development. Without exception, they all received advanced university training, and more than half of them had the opportunity to pursue postdoctoral research work.

In the meantime, the establishment of women's colleges significantly increased the participation of women in higher education and subsequently in

**Table 2. Origins of Undergraduate, Graduate, & Postdoctoral Training**

| Name | Undergraduate Institution | Undergraduate Field | Undergraduate Degree & Year Awarded | Graduate Institution | Graduate Field | Graduate Year Awarded | Postdoctoral Institution(s) |
|---|---|---|---|---|---|---|---|
| Marie Curie | Sorbonne (U Paris) | Physics | licenciée en physiques 1893 | Sorbonne (U Paris); Sorbonne (U Paris) | Mathematics | licenciée en sciences mathematiques 1894; Doctor of Physical Science 1903 | |
| Irene Joliot-Curie | Collège Sévigné, Paris | | baccalaureate 1914 | Sorbonne (U Paris); Sorbonne (U Paris) | Physics & Mathematics | licenciée 1920; D.Sc. 1925 | Radium Institute |
| Margaret Mead | Barnard College, New York | Psychology | B.A. 1923 | Columbia; Columbia U, New York | Psychology Anthropology | M.A. 1924; Ph.D. 1929 | |
| Barbara McClintock | Cornell U, New York | | B.S. 1923 | Cornell; Cornell U, New York | | M.A. 1925; Ph.D. 1928 | Cornell; U of Missouri, Caltech |
| Maria Goeppert-Mayer | U Göttingen, Germany | Mathematics | | U Göttingen, Germany | Physics | Ph.D. 1930 | Johns Hopkins U, Maryland |
| Rachel Carson | Pennsylvania College for Women (later named Chatham College) | English (minor in Biology) | A.B. 1929 | Johns Hopkins | Marine Zoology | M.A. 1932 | |

**Table 2.** (*Continued*)

| Name | Undergraduate | | | Graduate | | | Postdoctoral Institution(s) |
|------|---------------|-------|------------------------|-------------|-------|-------------|------------------------------|
| | Institution | Field | Degree & Year Awarded | Institution | Field | Year Awarded | |
| Rita Levi-Montalcini | | | | Turin U, Italy | Medicine | M.D. 1936 | Washington U, St. Louis, Missouri |
| Dorothy Hodgkin | Somerville College, Oxford | Chemistry | B.A. 1931 | Cambridge | | Ph.D. 1937 | Cambridge |
| Rosalyn Yalow | CUNY Hunter College | Physics & Chemistry | A.B. 1941 | U Illinois U Illinois | Physics Physics | M.S. 1942 Ph.D. 1945 | |
| Fay Ajzenberg-Selove | Michigan | Engineering & Physics | B.S. 1946 | Wisconsin | Physics | Ph.D. 1952 | Caltech & New York U |

science. Despite a lack of funds and other kinds of support for research, these institutions are dedicated to maintaining the highest possible academic standards among students (Ferry 1998:41; Pycior et al. 1996:23–24; Rossiter 1982, 1995; Tripp-Knowles 1995:29). Women's colleges have become places where students can see and seek out female role models in science. This was particularly important for female students when women were still invisible in education and employment. These institutions are also a magnet for women with scientific talents. Due to employment barriers, many women with doctoral training in science could not find academic jobs at major research universities. Therefore, women's colleges became a major employer of female Ph.D.s in science. In short, girls' schools and women's colleges are a primary producer and employer of female scientists.

Equally important, it is well documented that women's colleges constitute an ideal place where the first generation of female scientists can pick and groom their best students as successors (Lear 1998:45; Rossiter 1982:18–19). Contrary to common belief, there have been long traditions of *old-girl networks* and *invisible colleges* operated within women's colleges. Most female faculty of women's colleges would rather train their protégés than hire them from the outside. When a faculty member finds the right student, she offers that student support and encouragement. The faculty member eventually becomes the student's mentor and close friend, guides her protégé's selection of graduate schools, and pays close attention to her progress. Next, she would have the college hire the protégé back as a faculty member. Female faculty at women's colleges have used this strategy since the nineteenth century to create what Rossiter calls "protégé chains." Mary Skinker used a similar strategy to develop Rachel Carson's confidence and scientific talents (McCay 1993:7).

Indeed, some of the scientific pioneers on this list benefited from single-sex educational environments. Maria Goeppert-Mayer excelled in languages and mathematics at girls-only public elementary schools. Rosalyn Yalow had positive undergraduate experiences at Hunter College, at the time a highly competitive women's college in New York City, and graduated with high honors in chemistry and physics. Margaret Mead transferred to Barnard College after finding herself at odds with the student culture at DePauw University (Indiana). Mead preferred going to a women's college like Barnard for two reasons. For one, she could work as hard as she wanted to and pave the way for pursuing anthropology as a career. Another reason was that she did not have to compete with men in male-dominated fields.

Despite the potential benefits of partial segregation for female students and science Ph.D.s, these single-sex academic institutions may not always be suitable training grounds for female scientific pioneers. Others on the list found less-than-positive experiences in a single-sex learning environment. Based on the experiences of female pioneers in this study, these institutions could also be places that stifle creativity and scientific talent. Before coming to the United States and attending a girls' school (Julia Richman High School), Fay Ajzenberg-Selove attended a girls' school in Paris. While its discipline was firm, she found its teaching rigorous but dull, and the French school did not offer any science courses. College administrators at Pennsylvania College for Women (PCW) discouraged Rachel Carson and other students from pursuing science as an academic major or career choice. One of the reasons was the limited career opportunities for women with scientific training; female science graduates, even if hired, tended to hold teaching positions in high schools or lesser-known colleges. Additionally, PCW's president, Cora Helen Coolidge, firmly believed that women lacked the intellect or physical stamina for science careers (Lear 1998:43). Nonetheless, Carson graduated magna cum laude in 1929 with a B.A., majoring in English with a minor in Biology.

Taken together, these counterintuitive findings suggest that cultural transformations have not altered traditional gender role expectations. For the most part, the society still perceives new educational opportunities as just another tool to prepare women for their roles as wives and mothers. Especially for women in the midnineteenth and midtwentieth centuries, marriage and family still constituted an obligation, while pursuing a career in science was a matter of choices. Barbara McClintock's mother was initially opposed to her going to college at all, and was afraid that she might become a professor. Although Rita Levi-Montalcini graduated from a girls' high school in Turin with an excellent academic record, she was not academically prepared to enter a university. None of the subjects required for college entrance and scientific careers, such as mathematics, sciences, Greek, and Latin, were offered at her school. All this was the outcome of her father's plan—that his daughters (but not his only son) should attend a girls' high school and learn how to be good wives and mothers. Levi-Montalcini's father, Adamo Levi, attributed his aunts' unhappy marriages to overeducation, since these women earned their doctorates (in literature and mathematics) (McGrayne 1998:205).

In addition to the establishment of women's colleges, a proliferation of teachers' colleges and junior colleges in the midtwentieth century provided job opportunities for female scientists. During the 1940s, women accounted for half of all faculty at these institutions (Rossiter 1995:186). However, all subjects in this study launched their careers in coeducational institutions or government agencies. Their educational and career experiences suggest that girls' schools and women's colleges might play an important role in educating women, but single-sex institutions are far from being a crucial factor in the production and employment of female scientific pioneers.

Women's advancement in science is made possible when they have gate-keepers as their advocates. As power brokers, influential male scientists have control over the allocation of resources, rewards, and recognition. Once these men are convinced of women's significant value to the scientific community, they in turn could become a valuable source of sponsors, mentors, or role models for female scientists (Rayner-Canham and Rayner-Canham 1998). Clearly, all subjects in this study benefited immensely from the guidance and support of influential male figures at different stages of their lives and careers. As scientific spouses, teachers, peers, employers, colleagues, friends, or collaborators, these male scientists had a significant impact on the training and career development of female scientific pioneers. This assertion, however, does not diminish the contribution of women to the career development of subjects in this study. As noted in previous and forthcoming chapters, female scientific pioneers benefited enormously as well from the support offered by female relatives, friends, students, and teachers (e.g., Irene Joliot-Curie from Marie Curie, Margaret Mead from Ruth Benedict, Barbara McClintock from Harriet Creighton, Rachel Carson from Mary Skinker, Rosalyn Yalow from Edith Quimby).

During the nineteenth century, women succeeded in entering science by cooperating or collaborating with men. By doing primarily "women's work," they became valued members of a research team or organization (Rossiter 2003:63–65). Yet in many ways, these women occupied an important but subordinate role in research collaboration, and the male leader of the research team usually received the bulk of or all the credit for their joint work. This is not surprising, considering that male and female partners on the teams tended to have a paternal relationship due to differences in age and social status; these differences can contribute to and sustain gender division of labor and role differentiation in professional relationships. Nonetheless, according to Reskin (1978), the professional benefits of this type of relationship outweigh its costs for the female member of the team.

During the twentieth century, formal barriers for women to enter science were gone, and marriage to a scientist was no longer a major avenue for women to enter the scientific profession. Increasingly, female scientists enjoyed expanded opportunities for doing pioneering work through "professional marriages" instead of "legal marriages." As shown in this and subsequent chapters, having what Reskin (1978:13–16) calls a "professional spouse" is a major determinant of the career progress of talented female scientists. In fact, many female scientific pioneers in this study formed and maintained successful, productive collaborations with male scientists: (a) Marie and Pierre Curie, (b) Irene and Fred Joliot-Curie, (c) Margaret Mead and Gregory Bateson, (d) Maria Goeppert-Mayer and Joseph Mayer, (e) Rita Levi-Montalcini and Viktor Hamburger (and later Stanley Cohen), (f) Dorothy Hodgkin and Desmond Bernal, (g) Rosalyn Yalow and Solomon Berson (and later Eugene Straus), and (h) Fay Ajzenberg-Selove and Tom Lauritsen. Through gender division of labor and role differentiation, these scientific couples managed to accomplish a great deal in their joint careers that neither party might have achieved alone.

As discussed in the previous chapter, marriage has opposite effects on the scientific careers of men and women. By and large, being married is beneficial to the careers of male scientists, while it tends to be detrimental to the careers of female scientists. Conversely, married female scientific pioneers in this study who had a spouse who was also pursuing a science career found that alliance a real asset. All seven of them derived personal and professional benefits from "scientific marriages" (i.e., Marie Curie, Irene Joliot-Curie, Maria Goeppert-Mayer, Dorothy Hodgkin, Margaret Mead, Rosalyn Yalow, and Fay Ajzenberg-Selove). Somehow, the male and female partners found ways to make their dual-career marriages work for both of them.

The career choices of some female scientific pioneers lend credence to the "women's work" thesis. Based on Margaret Mead's accounts, she avoided direct competition with men in male-dominated fields by going into anthropology. As a cultural anthropologist, she concentrated and focused on the kinds of work that, in her opinion, women did better than men. Mead also succeeded in turning the disadvantages of being an elderly woman into advantages by: (a) working with children and women in situations where male participation is highly suspected and resented by the men of a society, and (b) working with men and women as an older woman: "using a woman's postmenopausal high status to achieve an understanding of the different parts of a culture, particularly in those cultures in which women past the reproductive period are freed from the constraints and taboos that constrict the lives of younger women" (Mead 1972:100).

As a woman and a newcomer in anthropology, Mead found her niche. Rachel Carson also successfully combined a feminist task (writing) with a masculine endeavor (science) as a career. Dorothy Hodgkin excelled in crystallography—a field which is relatively receptive to women's participation (Rayner-Canham and Rayner-Canham 1998:67–68). Despite the diverse backgrounds of female scientific pioneers in this study, like other women in science, their success might have been limited by performing women's work or concentrating in women's fields. This issue will be explored in detail in the next section.

## THE ECONOMIC FACTOR

Economic changes have given great impetus to women's advancement in science. Crucial support came when female scientists most needed it for sustaining their research or moving ahead. This vital support has promoted women's entry to science. Its direct impact on female scientists ranges from enhancing women's participation to promoting their career development by providing necessary startup funds or facilities for building labs or doing research, and fieldwork. The indirect impacts on the scientific community and society include:

1. Changing gender roles in society because of increased women's participation in education and employment.
2. Changes in the gender composition of the science workforce.
3. Increases in competition for resources, rewards, and recognition.
4. Diversification of the top of the scientific establishment.
5. Testing of the scientific establishment's normative and reward structures.

As shown below, female scientific pioneers have derived tangible benefits from economic development, including being able to stay in the science pipeline, gaining opportunities to conduct experimental studies or fieldwork, and earning opportunities to make significant contributions or innovations.

Since the late nineteenth century, private financial support has enhanced women's participation in higher education, graduate studies, and postdoctoral training. The private sector has taken initiatives to improve the recruitment and retention of female students and faculty in science. "Creative philanthropy" in the form of scholarships, postdoctoral fellowships, endowed professorships, and free dormitories and research facilities had a notable positive impact on their participation in science training and employment. "Coercive philanthropy" through threats of withdrawing support has also been used to advance the higher education of women (Rossiter 1982:47; 1995:38, 71).

Financial support for training and research became critical for female scientific pioneers in this book at different stages of their education and careers. For instance, a Polish female mathematician helped Marie Curie secure an Alexandrovitch Scholarship for the academic year 1893–94 when she was pursuing a degree in mathematics at the Sorbonne. This external support provided a crucial, if temporary relief to her financial problems. During the 1920s and 1930s, with support from the U.S. Radium Fund and other private contributions, Curie was able to equip the Radium Institute with additional radioactive materials and the latest scientific equipment.

After Margaret Mead's father withdrew his financial support, her fieldwork was supported by research and editorial assistantships as well as funds from sources such as the National Research Council (NRC), Columbia University, Social Science Research Council, museums, and later on, royalties from her books.

Barbara McClintock used grants from NRC, the Rockefeller Foundation, and the Guggenheim Foundation to work at a number of places including Cornell, Caltech, and the University of Missouri. She stressed the significance of this support on scientists' careers:

> For the young person, fellowships are of the greatest importance. The freedom they allow for concentrated study and research cannot be duplicated by any other known method. They come at a time when one's energies are greatest and when one's courage and capacity to enter new fields and utilize techniques are at their height (McGrayne 1998:157).

As for Rachel Carson, her undergraduate and graduate education was supported by scholarships from her state of residence, the PCW, and Johns Hopkins. Like McClintock, a 1951 Guggenheim Fellowship allowed Carson to support writing a guide to American seashores.

At the University of Michigan, Fay Ajzenberg-Selove was admitted into an extraordinary dormitory donated by Martha Cook, an alumnus. The congenial living atmosphere allowed her to blossom socially and intellectually. Again, a Guggenheim Fellowship allowed her to take a leave from Haverford and to do research at the Lawrence Berkeley Laboratory. In addition to having access to the excellent library and first-rate computer center there, Ajzenberg-Selove met many fascinating people. Most important of all, "it led to a reawakening of [her] joy in doing intense research in nuclear physics. . . . [it] was also a great boost to [her] morale" (Ajzenberg-Selove 1994:131).

New professional career opportunities became available to women in wartime. During the First World War, women constituted a growing number of federal clerical workers. Shortages of scientists, university teachers, and government workers during the Second World War presented educated women an opportunity to make inroads in the scientific professions. Women with advanced scientific training also took advantage of career opportunities in nonprofit organizations as well as in self-employment (McGrayne 1998:7; Rossiter 1982:219; 1995:235, 277). Marie Curie had taught at a normal school in Paris. The appointment as Director of the Red Cross Radiology Service in the early part of the twentieth century allowed her to combine her leadership skills and scientific knowledge to build radiology vehicles for the examination of millions of wounded soldiers. Her first radiological assistant was daughter Irene Joliot-Curie. Joliot-Curie, along with her mother, trained hundreds of radiological technicians to operate the medical facilities.

Later on, Joliot-Curie's appointment as the French Undersecretary of State for Scientific Research was the first cabinet post in France to link science with national development. Though Joliot-Curie had served for only six months, she accepted this key post in order "to make it easier for other women to enter the government" (Crossfield 1997:117). This post offered Joliot-Curie the opportunity and responsibility to consolidate scientific research in France.

During the Second World War, Margaret Mead worked for a committee of the National Research Council as an anthropologist.

Harold Urey, a 1934 Nobel laureate, recruited Goeppert-Mayer in the spring of 1942 to work on the Manhattan Project, in which the atomic bomb was being developed in secret. Not only did she receive a salary, she enjoyed the work.

Rachel Carson was able to combine her career interests in professional writing and marine biology in a government post. From 1936 to 1952, she made steady progress in rank and grade at the U.S. Bureau of Fisheries and subsequently at the Fish and Wildlife Service.

During the summer of 1947, Rosalyn Yalow joined the Bronx Veterans Administration Hospital, initially as a part-time consultant, when funds became available for establishing a radioactivity service. Several years later, she became a full-time physicist and chief of the radioisotope service at the VA Hospital. Working there also gave her the opportunity to meet Solomon Berson, whom she collaborated with for 22 years.

Career experiences of female scientific pioneers provide support for the proposition of sex-typing of skills (Reskin 1978; Wajcman 1995:191; Williams 1995). Unlike other female scientific pioneers in this study, Mead, McClintock, Carson and Yalow were able to develop their scientific careers beyond the confines of academe. Yet despite the progress women have made in science, many of the positions held by female scientists are less prestigious than those offered to their male colleagues—jobs that tend to have separate career ladders or are simply women's work. For example, women gravitate toward certain departments, such as schools of home economics and institutes of child welfare. In the field of psychology, women are more likely to be found doing clinical work while men occupy academic or administrative posts. Academic female scientists are more likely than their male counterparts to be research associates or research assistants to men, or to be promoted to positions such as dean of women instead of other administrative ranks (Nidiffer 2000; Rossiter 1982:203–205). Helson also found that, compared to creative female mathematicians, creative male mathematicians published more papers and held important positions at prestigious universities. A sizeable number of Helson's female subjects had no regular positions. Since most of the married were married to mathematicians, antinepotism was a problem (Helson 1971:216).

The career experiences of female scientific pioneers echoed the theme of sex-typing in occupations. When new career opportunities arose for the Curies, Marie Curie was offered a research position in a lab and later on, a lectureship, while her husband Pierre was made an assistant professor and subsequently a chair of physics.

When collaborating with Solomon Berson at the VA Hospital, Rosalyn Yalow did women's work. As an administrator for the research team, her tasks ranged from making airline reservations, to arranging for manuscript typing, to setting up experiments. Although Yalow and Berson considered each other equals, Yalow managed the lab while Berson was the "front man" outside their lab. He was the one who traveled to meetings to give talks to and deal with the scientific establishment (McGrayne 1998:343–344). In short, it was Berson, not Yalow, who did the "networking" and dealt with the power brokers.

Maria Goeppert-Mayer's professional experiences provide classic examples of how antinepotism rules at universities have affected women's careers in science. To avoid potential conflicts of interest, there was an explicit rule prohibiting the hiring of married couples in the same department or institution. Generally, the wife was offered a position and salary, if any at all, that was not commensurate with her qualifications and experience (Pycior et al. 1996; Rossiter 1982:142, 1995:122). During the 1950s and 1960s, young female scientists tended to have nontenure appointments as instructors, lecturers, or research associates, while their scientific spouses held tenure-track academic positions as assistant professors (Pycior et al. 1996:32, 56, 76). As a result, many married female scientists experienced "title deflation" during most of their careers. For decades, Goeppert-Mayer accepted part-time teaching or research positions at universities with little or no pay. In 1946, when the Mayers moved to Chicago, her husband Joseph was offered a full professor position, while Goeppert-Mayer was an associate professor. She did not receive the offer of a full professor's salary from Chicago until 1959, when the University of California in La Jolla offered both Mayers full professorships.

Because few full-time teaching positions were available to women in 1935, Rachel Carson resorted to writing as her primary source of earnings. Carson ran into a similar situation one decade later when she wanted to move from the government to the private sector. To maximize her chances of success, Carson emphasized her abilities as a writer and downplayed her expertise in marine biology. Even though her strategy was consistent with traditional gender-role expectations, she was unable to find professional work outside the government. At that time, none of the conservation and zoological organizations had women on their professional staffs. This outcome is consistent with advice from the president of the Pennsylvania College for Women, Cora Helen Coolidge, and her mentor, Elmer Higgins of the U.S. Bureau of Fisheries: (a) there were very few scientific jobs for women, (b) a woman would never get a job in industry, and (c) she would have to make a living in teaching or government work.

During the 1940s and 1950s, most U.S. universities would not hire women as faculty. Ajzenberg-Selove could not find teaching positions at major universities. Initially, she had taken one-year teaching positions in places such as the Navy Pier branch of the University of Illinois (a community college in Chicago) and Smith College. She also held appointments of assistant professor at Boston University, visiting fellow at MIT, and research professor without tenure at the University of Pennsylvania. At Haverford College, she rose to the rank of tenured full professor. But Ajzenberg-Selove soon realized that her career in physics would not blossom at a men's college because of its heavy teaching load as well as its emphasis on teaching rather than research and scholarship.

## THE POLITICAL FACTOR

Major political developments also had a notable, positive impact on women's participation in education and employment. In the United States, a series of legislative changes introduced in the 1960s and 1970s allowed female scientists to play a bigger role in shaping the education and employment of women. For instance, the passage of the Civil Rights Act in 1964 prohibited discrimination in education and employment based on sex. The Equal Employment Opportunity Act of 1972 further loosened social constraints on women's participation in the labor force. According to Rossiter, a legal revolution in women's education and employment rights took place between 1968 and 1972. During the same period, she observed a women's movement in science and engineering (Rossiter 1995:361–382). Female scientists, collectively and individually, took an increasingly active role in enhancing women's status in schools, professional societies, and workplaces.

Marie Curie was a beneficiary of "scientific enthusiasm" in Europe. Emphasis on scientific and technological progress, along with electricity's transformation of industry and communications, created a sense of hope and excitement. As part of this subculture, Curie joined a "Flying University" where a group of friends taught each other science, technology, and politics. This collective effort to improve women's status in science signified the rise of a women's movement in science.

Some scholars have attributed Marie Curie's receipt of the second Nobel Prize for chemistry in 1911 in part to political factors. Her personal and political circumstances make it impossible, however, to separate the effects of "merits" and "politics" that are associated naturally. This lingering doubt is summarized succinctly by Moulin (1955:259):

> Without desiring to minimize her merit, which appears to have been very great, one may perhaps wonder to what extent certain factors outside those of pure science—the active pro-Polish sentiment and feminism at the beginning of the century and also the death of Pierre Curie in tragic circumstances—weighed in her favor.

Modis (1988) also found that the increase in the number of female Nobel Prize recipients coincided with the rise of feminism during the 1930s and the late 1970s to early 1980s. Women's progress in science is dependent in large part on their status in the society. In fact, with the exception of the Curies, all female Nobel laureates in science received their prize after the Second World War. An optimistic reading of this data suggests that the scientific establishment has begun to bestow significant recognition and honor on women in the last half of the twentieth century, an indication of women's progress in science.

The First World War gave Marie Curie an opportunity to work closely with her daughter Irene Joliot-Curie, initially at the front and subsequently at the Radium Institute. As a result, Joliot-Curie became her mother's favorite research assistant, collaborator, and successor. Political upheavals in France offered Joliot-Curie a chance to enter politics as a female scientist. For the first time, politicians were eager to bring women into the government. Joliot-Curie was one of three women politicians considered progressive in outlook and universally known. As the Undersecretary of State for Scientific Research, she served as a role model for women in France and elsewhere.

Margaret Mead's findings on human nature and different cultures resonate with the political trends of her time. The period between 1920 and 1950 was an era of unprecedented changes in the United States and around the world, and people were receptive to new and radical ideas. Mead's anthropological work and social-policy recommendations found growing support among the middle class and early viewers of the television era.

Due to a unique set of changing circumstances, Barbara McClintock survived and eventually thrived outside the confines of a university setting. Her unpleasant experiences at a number of academic institutions clearly indicated that a tenured university position, which would require her to fulfill regular teaching and administrative obligations, was out of the question. On the other hand, as a research associate at Cold Spring Harbor, she enjoyed the freedom to continue her research on maize in a scientifically stimulating environment. As Keller described her situation: "[Cold Spring Harbor offered Barbara McClintock] an environment in which she could pursue her own ideas and do her work as she saw fit, protected from departmental politics, from teaching duties, from administrative responsibilities" (1983:109).

All this was made possible at a time when political establishments in Washington had become more receptive to women's participation in science. Despite her having a reputation of idiosyncrasies, the financial support she received from various funding agencies is a testimony to the argument of the changing political climate for women's progress in science.

Female scientists from abroad have also benefited from political changes in the United States. Their scientific careers blossomed in their host country. In 1930 at the age of 24, Maria Goeppert-Mayer earned her doctorate in theoretical physics at the University of Gottingen, but she spent her entire postdoctoral career in the United States. It was difficult for women to become professors in Germany. For this reason, both Goeppert-Mayer and her mother felt strongly that her scientific career should develop in a different political context, such as the United States.

Rita Levi-Montalcini felt exactly the same way after spending 26 years at Washington University in St. Louis. Those were "the happiest and most productive years of my life. . . . I felt at home the day I landed. There is great cordiality, generosity . . . And America is a society in which merit is genuinely rewarded—you cannot say the same for Italy" (McGrayne 1998:212).

Dorothy Hodgkin's scientific career took off for a number of political reasons. Women enjoyed a high visibility in crystallography at the time (Rayner-Canham and Rayner-Canham 1998:67–91). Equally importantly, the postwar era was a golden period for scientists in Great Britain, where a renewed emphasis on scientific and technological progress coincided with the proliferation of funds for research activities. Hodgkin's career benefited from women's improved status in the field as well as the availability of funds for scientific research.

The drafting of young men into the military during the Second World War allowed Rosalyn Yalow and other women to gain entry into graduate school. There were few assistantships available at that time, especially for Jewish women. Still, due to a shortage of male graduate students, Yalow was able to obtain an assistantship at the University of Illinois at Urbana-Champaign. Her work and career took off at the Bronx VA Hospital, the national flagship hospital for the emerging VA system as well as a major center for medical research and training.

The trajectory of Fay Ajzenberg-Selove's career must be understood against the backdrop of a series of political developments. The post-WWII period was an exciting time for physics. New fields had proliferated at Caltech, and public funds for scientific research were readily available. Her self-nomination for a tenured position at the University of Pennsylvania was voted down by the department. However, with the support of key female advocates for women faculty, Ajzenberg-Selove eventually won her legal case over the faculty appointment with the Equal Employment Opportunity Commission and became professor of physics. The inspiration and support received throughout her career also came from her active involvement in the establishment of mainstream scientific-association branches.

Other researchers have also underscored the importance of changing political circumstances on women's improved status in science. For example, Levin and Stephan (1998) noted that wage compression as well as affirmative action during the 1970s, among other factors, resulted in university employers granting women larger-percentage salary increases than men.

## IMPLICATIONS

It is important to consider how some of the positive and negative historical forces influenced the education and employment of women in science. Until recent decades, women occupied a marginal role in the development of science. Participation as well as advancement in science was not simply an individual matter, however. Social and legal circumstances restricted women from formally participating in scientific activities.

On the other hand, changing dynamics in the society opened up possibilities for women to *develop* their scientific talents. Transformations in culture, the economy, and politics have allowed women to obtain advanced training in science and to work side by side with their male counterparts in research. Expressions such as "being at the right place at the right time," "contextual factors," or "luck" have been used to characterize these structural transformations (Murray 2000:229; Rosenbaum 1984:274; Stephan and Levin 1992:90–91). Regardless of the terminology used, all subjects in this study took advantage of various opportunities in the twentieth century to pursue pioneering work.

It is imperative to rethink mobility in science from a sociological perspective. Science careers are socially situated. Women's entry into the community of scientific elites may bear semblance to processes of diversifying the power elites as well as the trustees of elite arts boards (Hermanowicz 1998; Ostrower 2002; Zweigenhaft and Domhoff 1998). This means that it is necessary for the society to allow for women's greater participation in science, and certain societal barriers for women in education and employment must be dismantled. Additionally, the scientific community must adapt and change for renewal and growth: to make adjustments in its structural arrangements to accommodate and assimilate newcomers. One such change is that opportunities to study and participate in science fields become available and accessible to women.

Equally important, virtually all subjects in this study, with the exception of Fay Ajzenberg-Selove, seemed to have spent little or no time on solving or fighting many of the structural problems women faced in science. It could be argued that a few of them gave symbolic support to improving women's progress in science by participating in political or national organizations for (university) women (e.g., Irene Joliot-Curie, Margaret Mead, Barbara McClintock, Rachel Carson). But they spent little effort in making academia or science a better place for those who came after them.

Overall, there is more to the making of scientific pioneers than talent and opportunities. To understand how and why some but not all women with the right qualities and good timing have managed to reach the top of the scientific establishment, we need to move from the *development* to the *advancement* of female scientific pioneers. We now turn our attention away from the structural opportunities (i.e., social control or lack of) to the institutional forces (i.e., the environment) that are conducive to doing pioneering work.

## Chapter Five

# Reaching the Top: Institutional Forces

## INTRODUCTION

While a positive trend, the increase of women's presence in science and employment is far from challenging the traditional hierarchy in the scientific establishment. To join the ranks of scientific elites, it is not enough simply to get one's foot in the door. To do pioneering work and to be recognized for one's own ideas, more is required than mere structural opportunities. Additionally, the success of a female scientific pioneer cannot be credited to a single individual. What are the ahistorical effects that enabled all ten notable women in this study to succeed in science? I argue that forces of selection help facilitate talented women to reach the top of their fields: An individual has to be surrounded and supported by some extraordinary people, and pioneers-to-be work hard and take advantage of what the family, institutions, and society provide.

## FAMILY

Family is one of the most important agents of social change. Studies of the eminent have revealed *no* specific patterns of family that could produce "genius" (Goertzel and Goertzel 1962; Goertzel, Goertzel, and Goertzel 1978). Yet familial influence on female scientific pioneers in this study was quite strong, and manifested in many aspects of their lives and careers. By listening, coaching, and supporting, the (grand)parents of female scientific pioneers made lasting impressions on their lives. Their (grand)parents did not always tell them what they were supposed to do. However, the female scientific pioneers actually learned what and how to do the right thing in life by watching them. As children, they might not always have paid attention to what their (grand)parents were saying, but they were always watching.

These female scientific pioneers' families shared four commonalities:

*Most came from middle-class, academic, or professional backgrounds.* Typically, ancestors of their (paternal or maternal grand)parents had extensive business or professional experiences. Either one or both of their parents had received formal schooling or professional training. Growing up in a family with a tradition of learning or professional experiences gave these women direct exposure to positive role models early on.

Marie Curie was brought up in an impoverished-but-professional home. Her parents were members of the Polish urban intelligentsia with roots in the small-landed gentry. Her father taught physics at a boys' secondary school in Warsaw. Until Curie was born, her mother was the principal of a private girls' school.

Irene Joliot-Curie came from a relatively privileged background. Not all female scientific pioneers had the good fortune to be raised by parents who had won the Nobel Prize, and Joliot-Curie was the luckiest one; she was born into a literal scientific dynasty. Both her grandfathers were highly educated, and she was the eldest daughter of a very famous Nobel Prize-winning couple, Pierre and Marie Curie. She was inarguably raised in a scientifically stimulating milieu. One could use Joliot-Curie as a perfect example of the significance of an *extremely* supportive environment on career aspirations and career development.

Margaret Mead, brought up in a middle-class professional family, was the third generation of professionals in her family. Her paternal grandparents were teachers, while her maternal grandparents were well-traveled seed merchants and community leaders. Margaret's father, who had a doctorate in economics, was an economics professor at the University of Pennsylvania's Business School; her mother was a graduate student in sociology and carried on her studies while raising four children.

Barbara McClintock's father was a medical doctor. Her mother, an amateur painter and poet, gave piano lessons to help with family finances when needed.

Maria Goeppert-Mayer was the only child in a well-educated upper-middle-class family. On her father's side, she was the seventh generation of a family of university professors. Her father was a professor of pediatrics at the University of Gottingen in Germany, and her mother taught French and piano before marriage.

Rachel Carson's family was not well-off, but was land-rich. Her father tried unsuccessfully to combine farming with real estate. Rachel's well-educated mother, a gifted musician, gave up teaching after marriage.

Rita Levi-Montalcini grew up in an educated middle-class family. Her father was an engineer by training.

Dorothy Hodgkin's parents came from conventional well-to-do families. Her grandfather studied classics at Oxford before serving as a missionary in India. Both of her parents were archaeologists.

Of all ten female scientific pioneers, Rosalyn Yalow had the most humble beginnings. She grew up in a poor, uneducated family. Her father quit school after eighth grade and became a streetcar conductor before opening a paper-and-twine business. Her mother left school after sixth grade and did garment piecework at home.

Fay Ajzenberg-Selove's father was a mining engineer. Before going bankrupt during the German Depression, he was the director of a sugar refinery in Lacinate (20 miles south of Paris). Her mother was a high school graduate.

*All subjects in this study grew up with a family tradition of emphasis on education for acquiring moral and practical training.* Many female scientific pioneers were raised in intellectually or scientifically stimulating environments. Their parents also viewed education as a means of developing self-reliance for daughters.

Marie Curie's parents took the education of their children very seriously. They expected their children to be able to read before entering school. After her mother's death, Curie's father continued to instill his love of science and literature into his children. Curie passed on this intense intellectual spirit to her two daughters. Instead of sending Irene and Eve Curie to public schools for formal education, she and her colleagues set up a private cooperative (secondary school) for their children. Irene Joliot-Curie, along with a group of young boys and girls, learned every day from the most distinguished scholars and scientists in their fields. For Joliot-Curie, mastery of algebra and calculus was considered as normal as swimming. Despite skepticism, the school's unconventional approach to educating youngsters succeeded in preparing many of its graduates, including Joliot-Curie, for scientific careers (Crossfield 1997:102). To some extent, Joliot-Curie was following her father's example. Pierre Curie was taught at home by his father, Eugene Curie, in a similar fashion.

The Mead family was also revolutionary when it came to educating their children. Frequent relocation of her family precluded Margaret Mead and her younger siblings from attending schools regularly until high school. As a result, her mother resorted to hiring local experts to teach her children whatever they knew best. Mead received lessons in traditional academic subjects from her grandmother at home. Familial influence on Mead's love for learning is clear: "All my life I expected to go to college and I was prepared to enjoy it. For me, not to go to college was, in a sense, not to become a full human being" (Mead 1972:88).

On balance, the McClintocks played a strong but less direct role in their daughter's education. Barbara McClintock's mother initially had reservations about her daughter attending college, because she was afraid she would become a professor. Finally, she yielded to her husband's decision that Barbara McClintock would go to Cornell.

Like Irene Joliot-Curie, Maria Goeppert-Mayer was brought up in an intellectually vibrant setting. Aside from being the only child in a well-educated family, many close family friends were prominent mathematicians and physicists. Finally, Goeppert-Mayer's father expected her to receive a good education and to be eventually economically independent.

Recognizing their daughter's aptitude for learning, the Carsons sent Rachel Carson to college despite financial hardships. Her mother tutored Carson at home and instilled a love of nature in her.

Rita Levi-Montalcini's mother fully supported her daughter's decision to study medicine. Despite his early reservations, Levi-Montalcini's father hired tutors to prepare her for university entrance exams.

Dorothy Hodgkin's upbringing bore some semblance to Margaret Mead's. Family dislocations and separations from her parents were a fact in Hodgkin's childhood and adolescence. However, her parents also valued education above everything else. Since she was the oldest of three girls in the family, Dorothy Hodgkin's father expected her to be educated the same way as a son. Her mother, whenever possible, tutored her daughters at home. To cultivate her interest in science, her mother gave Hodgkin a children's book by William Henry Bragg, among others. Bragg and his son, William Lawrence Bragg, were corecipients of the 1915 Nobel Prize in physics for their contribution to the study of crystal structures by x-rays. As leaders of x-ray crystallography, the Braggs trained scores of female scientists in the field (Rayner-Canham and Rayner-Canham 1998:68–69). William H. Bragg's book turned out to be an inspiration when Hodgkin grew up to become a first-rate x-ray researcher. Further, her father took the responsibility of negotiating the complexities of her entry to Oxford.

Although Rosalyn Yalow's parents had humble origins, they both valued learning. Her father read *The New York Times* and her mother read every schoolbook Yalow and her elder brother, Allie, brought home from school. Allie had taught Rosalyn to read before she began kindergarten. Before turning five, Yalow, along with her brother, paid weekly visits to the local public library and eventually became members.

*A male or female figure in the family was instrumental in nurturing their interest in the pursuit of knowledge.* Typically, a parent, grandparent, or close relative played an important part in their development as a scientist. Many of them identified closely with their father or another male figure, such as their grandfather. This is consistent with the observations of female mathematicians by Helson (1971) and Maccoby (1970:22). Family members also provided essential emotional and/or financial support for their education, and subsequently their pursuit of scientific careers.

Besides parental influence, Marie Curie had an extremely close relationship with her sister, Bronya. Both sustained each other's lifelong interest in learning and in career pursuits by offering mutual support. After the death of Marie Curie's mother-in-law, her father-in-law, Dr. Eugene Curie, moved into her household and assumed responsibility for his granddaughters, Irene and Eve. His availability and willingness to help contributed to Curie's freedom to study and research seven days a week. However, of all the family members, Curie regarded her father, Wladyslaw Sklodowski, as the most important figure in her life:

> My beloved father must stop despairing about not being able to help us. It is inconceivable that my father could do more for us than he has done. We have a good education, a solid cultural background. . . . We will make out all right, without doubt. As for me, I will be eternally grateful to my dear father for what he has done for me, because he has done so much (Quinn 1995:74).

Without a doubt, her widowed mother, Marie Curie, was *the* most critical person in Irene Joliot-Curie's career. Their work and life intertwined with one another's, especially after the death of Pierre Curie. Another important figure in Joliot-Curie's moral and intellectual development was her widowed grandfather. Eugene Curie helped and influenced the young Irene, and raised and mentored Joliot-Curie until his death in 1910.

Like Irene Joliot-Curie, in addition to her parents, Margaret Mead underscored her paternal grandmother's importance on her personal and professional development. Mead learned poetry, arithmetic, and botany from her grandmother, Martha Ramsey Mead, a college graduate as well as a former teacher and school principal. From her, Mead also learned the importance of finishing her tasks as well as taking responsibility for her younger siblings when her mother was away. Her influence on Mead was profound, evidenced by Mead's recollection:

> My paternal grandmother . . . was the most decisive influence in my life . . . She became my role model when, in later life, I tried to formulate a role for the modern parent who can no longer exact obedience merely by virtue of being a parent and yet must be able to get obedience when it is necessary (Mead 1972:45–46).

When Mead's father opposed her attending college (because he was afraid that his daughter would marry rather than pursue a career), her mother, Emily

Fogg Mead, stood by Mead and came up with idea of sending Mead to her husband's alma mater, DePauw University in Indiana, instead of hers, Wellesley College. Edward Sherwood Mead compromised and let Margaret attend De-Pauw. Her father also funded Mead's maiden voyage to the South Pacific to do fieldwork. Mead acknowledged his enduring influence on her self-development:

> It was my father, even more than my mother, whose career was limited by the number of her children and her health, who defined for me my place in the world. Although I have acted on a wider stage than either my mother or my father, it is still the same stage—the same world, only with wider dimension (Mead 1972:44).

Unlike other female scientific pioneers, Barbara McClintock did not think that her parents had directly supported her intellectual pursuits. In fact, she did not consider them, or anyone else, as role models. On the other hand, she could not deny the backing of her father, Thomas Henry McClintock: "I just knew what I wanted to do. It was easy because it was so clear and because I had the support of my father, the complete support" (McGrayne 1998:149).

Early in life, Maria Goeppert-Mayer's father inspired and encouraged her to become a scientist as well as the seventh-generation university professor in the family. Friedrich Goeppert insisted that his daughter "never become just a woman."

One can hardly overstate the influence of Maria Carson on her daughter's development into a marine biologist and science writer. Mrs. Carson not only nurtured Rachel's interest in nature, wildlife, and reading, she also encouraged her daughter's writing ambition, financed her college education, and later helped manage her sciencewriting projects.

Both Margaret Mead and Rita Levi-Montalcini had dominant father figures who played a crucial role in their lives and careers. They also had mothers who were exceptionally supportive of their academic and career plans. It was Adele Montalcini who urged Levi-Montalcini to accept eminent embryologist Viktor Hamburger's invitation to go to Washington University in St. Louis and stay there. Levi-Montalcini added her mother's maiden name, Montalcini, to her father's surname to distinguish herself from the other Levis (McGrayne 1998:202). Similar to Marie Curie and Margaret Mead, Levi-Montalcini admitted her father's enduring influence on personal development:

> It was [my father] rather than [my mother] who had a decisive influence on the course of my life, both by transmitting to me a part of his genes and eliciting my admiration for his tenacity, energy, and ingenuity. . . . From him, I inherited seriousness and dedication to work, a secular Spinosa conception of life (Levi-Montalcini 1988:16–17).

As mentioned earlier, as a result of her father's plan and action, Dorothy Hodgkin was able to enter Oxford's Somerville College to study chemistry in 1928. Her mother taught Hodgkin and her sisters all the subjects she knew at home. To encourage her children to find answers through systematic investigations, Molly Crowfoot encouraged Hodgkin to conduct her own experiments in the lab set up at the house.

Fay Ajzenberg-Selove also attributed her interest and success in science to her father's influence:

> I adored my father. . . . grew up following in his footsteps, striving to like the things he liked. . . . I walked with [my father] around the factory, enjoying the beauty of the machinery, smelling the sugar pulp, and watching his confident and comfortable discussions with the workers (Ajzenberg-Selove 1994:7).

*Their upbringings did not correspond to the gender role expectations at the time.* As noted in Chapter 2, women's underrepresentation in science has been attributed in part to early socialization of women into traditional gender roles. Conversely, these female scientific pioneers' early interest in science and persistence in doing scientific work could have been the outcome of their own (or parental) indifference to conventional gender roles. As noted in preceding sections, many of their parents (or close relatives) did not raise them according to the social expectations for males and females. Instead, through their actions and behavior, they expected their daughters to choose their subjects of study or careers following their own inclinations instead of the academic and career interests typical for women in society. In fact, for some female scientific pioneers, engaging in science was part of their childhood. Unlike most female students in science and engineering, these women already had a "tinkering" experience in childhood (McIlwee and Robinson 1992). Through their involvement in family science, these women saw firsthand the practice of science. More importantly, they learned early that, contrary to societal expectations, being a woman and learning science were compatible. This early exposure became an important mechanism in stimulating and sustaining their interest in science.

It could be argued that subjects in this study were "gender radicals," a term coined by Ortner to label individuals who question or break gender rules. According to Ortner, because of the masculine nature of mountaineering, women who began to participate in the sport in the 1970s were, in some sense, a "gender radical" (1999:217).

As noted by Marie Curie's brother, Josef, none of his three sisters—Helena, Bronya, and Marie—seemed to fit the role of a young girl typical for that time. Instead, they were all prepared to postpone marriage or remain single, intended to acquire higher education, and planned to have independent careers. Given the precedence of female independence in the family, their academic and career plans were not entirely unrealistic. Their mother was a girls' school principal. Uncle Zdzislaw's wife, Maria Rogowska, founded factories and ran the family estates. Aunt Wanda Sklodowska, the most educated woman in the family, attended college in Geneva and assumed a literary career (Quinn 1995:65).

Irene Joliot-Curie's parents did not have to "talk science" to her. She was born into and brought up in a scientifically stimulating milieu. Her life revolved around her parents' work and their lab. She studied and socialized with children of prominent scientists, many of whom were her parents' colleagues or close friends. She practically grew up at the Radium Institute. All this prepared her to have a lifelong pursuit of scientific research, and like her mother, Joliot-Curie intended to combine family and science in her life.

Margaret Mead's occupational interest and image contrast with the stereotypes of academic scholars for her time. The "girl in the neighborhood" was energetic, adventurous, and daring; traveled to distant places; and had "gone native" with people who were untouched by western civilizations. For the most part, Margaret Mead and her third husband, the British-born social anthropologist Gregory Bateson, felt they were deviants in their own cultures. Certainly, Mead did not fit the roles of an American homemaker or career woman. She was determined to have a professional career as well as children. Her aspirations probably came because she had two positive female role models in her life, which she described in the following way:

> It was my grandmother who gave me my ease in being a woman. . . . She had gone to college when this was a very unusual thing for a girl to do. . . . She had married, had a child, and she had a career of her own. All this was true of my mother as well. . . . The two women I knew best were mothers and had professional training. So I had no reason to doubt that brains were suitable for a woman. . . . Looking to my grandmother and my mother, I expected to be both a professional woman and a wife and mother (Mead 1972:53–81).

Margaret Mead had her first experience of fieldwork through her mother's work among the Italians living in Hammonton, New Jersey. From her grandmother, she learned to use both inductive and deductive approaches to solve problems:

> On some days, she gave me a set of plants to analyze; on others, she gave me a description and sent me out to the woods and meadows to collect examples. . . . I learned to observe the world around me and to note what I saw (Mead 1972:46).

Mead got her first experience of taking field notes when her grandmother asked her to take notes on the behavior of her youngest sisters Elizabeth and Priscilla. Of this, she related, "I learned to make these notes with love, carrying on what Mother had begun. . . . In many ways I thought of the babies as my children, whom I could observe and teach and cultivate" (Mead 1972:64). This training set the stage for developing her habit of taking extensive field notes as well as making careful observations in every fieldtrip throughout her career.

Mead's lifetime practice of writing up field notes in monograph form was also motivated by a determination to change the traditional practice in anthropology:

> During my student days I had been deeply impressed with the dreadful waste of fieldwork as anthropologists piled up handwritten notes that went untranscribed during their lifetime and that no one could read or work over after they died. . . . I vowed that I was not going to do this, that I would write up each trip in full before undertaking the next one (Mead 1972:183–184).

Mead's lively and allusive writing can be traced back to her poetry writing beginning at the age of nine. As she stated, "writing letters has been a very real part of my life, especially in the years I have been in the field" (Mead 1972:81). Her desire for autonomy and independence was demonstrated in her numerous expeditions to study people in different cultures: "I would be able to work in a world that had not been constructed for a woman to work in. I was beginning to realize that the freedom to work as one wished was the important thing" (Mead 1972:132). When doing fieldwork on remote islands, letter writing became her only means of staying connected with family and friends.

Barbara McClintock's parents changed her original name, "Eleanor," to "Barbara." They considered the former a feminine and delicate name and the latter a more masculine name, which seemed to fit their daughter better. For the most part, her parents raised McClintock as a boy. Her father gave her boxing gloves when she was four. Her mother did not stop her from ice skating, roller skating, bicycling, and playing basketball with boys. McClintock became mechanically inclined later in life due to her unconventional upbringing and early interest in machines and tools. When she lived with her relatives in Massachusetts as a toddler, her uncle taught her to repair machinery and nurtured her love for nature. Her father "tells her that at the age of five [she] asked for a set of tools" (Keller 1983:22). However, she was not exactly a tomboy. On the one hand, McClintock was not interested in playing with girls. On the other hand, she was not completely accepted by boys either. Consequently, she grew up free from any pressure to conform to traditional gender expectations. She was keenly aware of the advantage of being an exception to the norm: "My parents supported everything I wanted to do, even if it went against the mores of the women in the block. They would not let anybody interfere" (McGrayne 1998:148). The need for freedom and independence became a trademark of her scientific career. All this explains in part why she found it so difficult to meet the expectations of a scholar in the academe and why her male colleagues found it less-than-easy to find a professional home suitable for her.

Maria Goeppert-Mayer grew up in a university town populated by academics and prominent scientists. Her parents encouraged her to be curious and adventurous. Her father inspired his only child to use her mind and took her on science walks. At the age of seven, she watched a solar eclipse using the filters he made for her. More importantly, she adopted his low opinion of women and women's traditional roles. This is probably why she preferred the company of men throughout her life:

> My father said I should have been a boy. . . . He said, don't grow up to be a woman, and what he meant by that was a housewife . . . without any interest. I mean, he saw so many women who had just played with their children and had no interests whatsoever, and this he didn't like. . . . I felt flattered and decided that I wasn't going to be just a woman (Dash 1973:238).

Rachel Carson's success in combining science and writing as a career reflected her belief that a woman could and should have as independent a professional life as a man would. The experience of her gifted mother as a homemaker convinced Carson that she did not want to follow her mother's example. She had no interest in being a wife and a mother. Her choice of scientific writing as a career also ran counter to gender-role expectations at the time:

> People often seem to be surprised that a woman has written a book about the sea. This is especially true of men. Perhaps they have been accustomed to thinking of the more exciting fields of scientific knowledge as exclusively masculine domains. Then even if they accept my sex, some people are further surprised to find that I am not a tall, oversize, Amazon-type female (Lear 1998:214).

Like other female scientific pioneers, Carson had early exposure to science as a child: "I can remember no time when I wasn't interested in the out-of-doors and the whole world of nature. . . . Those interests I inherited from my mother and have always shared with her" (Lear 1998:8). At the age of ten, she entered a contest for science writers and won a silver badge. More important to her, perhaps, her piece was published. This first taste of literary success motivated Carson to write more stories in her adolescence.

As noted earlier, Rita Levi-Montalcini's father had reservations about sending her to college for fear of difficulty in combining family and career for a woman. However, like Rachel Carson, Levi-Montalcini did not want to be a wife and a mother. She had no interest in feminine pursuits, but showed a passion for trains as a child. Like Barbara McClintock, Levi-Montalcini's rejection of the traditional gender roles for women was unmistakable:

> My experience in childhood and adolescence of the subordinate role played by the female in a society run entirely by men had convinced me that I was not cut out to be a wife. Babies did not attract me, and I was altogether without the maternal sense so highly developed in small and adolescent girls (Levi-Montalcini 1988:35).

Contrary to Levi-Montalcini's belief that family and career are incompatible pursuits, Dorothy Hodgkin modeled after her parents and had both a marriage and a scientific career. Equally important, her experience in childhood and adolescence made it possible for her to choose science as a career. Conformity was not considered a virtue in her family. As a child, Hodgkin traveled to the countryside with her younger sisters, all by themselves, to explore nature. Early exposure to crystals coupled with frequent travel in adolescence expanded her horizon and sparked her interest in scientific research. After receiving a portable surveyor's box from Dr. A. F. Joseph during a visit of the Welcome Lab, Hodgkin used the tools to carry out simple chemical experiments in the attic lab at home.

Rosalyn Yalow also grew up in an environment that encouraged her to follow her own interests rather than those considered suitable for only women. As a child, she learned the virtue of fighting back. Her father, in particular, urged Yalow to do whatever boys did. However, Yalow did not reject the traditional gender role expectations for women. As mentioned in the preceding chapter, she held the belief that all female scientists should marry and have children.

Fay Ajzenberg-Selove's parents were instrumental in nurturing her scientific interests: "I had never been told that being curious—asking questions, seeking answers—and that being strong was not appropriate for a woman" (Ajzenberg-Selove 1994:55). Like most of the female scientific pioneers in this study, Ajzenberg-Selove enjoyed masculine pursuits:

> I built model airplanes . . . climbed onto the roof of a garage with a band of boys . . . solved mathematical and scientific problems . . . never had a doll . . . took flying lessons . . . liked shooting and became quite good at it [and] entered Michigan intending to become an aeronautical engineer and a pilot" (Ajzenberg-Selove 1994:7, 20, 48).

She admitted that her father and several female role models made it possible for her to circumvent the social constraints on women:

> The expectations of Misha [Dad] and the examples of my aunt Sara and my friend Lida, both independent women, were so much more important in my life than was the influence of my peers. I was too naive and unaware to be caught in sexual traps. . . . Being one of the boys seemed entirely satisfactory to me at the time (Ajzenberg-Selove 1994:55).

All this underscores the significance of childhood families, especially the parents and influential relatives, on the career aspirations of female scientific pioneers today. Analyzing different aspects of these women's familial influences tells us (a) why and how they chose science and (b) why pursuing it was important to them. Another noteworthy finding was that some female scientific pioneers encountered subtle or blatant bias, even within their own family and during their upbringing.

## SCHOOLS/INSTITUTES

Organizational support is critical in the career development of female scientific pioneers. As shown in Table 2 in the preceding chapter, the educational and career experiences of subjects in this study corroborate this assertion. Not all of them worked independently. They were not doing science in "mom and pop" shops. They were employed directly or indirectly by the private or public sector. It is also true that many of them lacked a power base in universities or research institutes. Nonetheless, many held what Rossiter called a voluntary professorship at a university or a position at a nonprofit institution (1995:331–332). This is not surprising, given that science is a social activity and that female scientists are relatively isolated in the scientific community. Institutional affiliation becomes a congenial way for women to establish and maintain (in)formal ties with the scientific establishment. At these institutions, women can surround themselves with people who are also committed to science and are willing to work with them to achieve their goals. Great scientific masters or advocates might attempt to select and groom them as future leaders in science. All this suggests that subjects in this study had recognized the significance of membership in the invisible college of their colleagues on scientific productivity and recognition (Crane 1965:701).

Besides the right networking opportunity, many scientists need lab space and assistance for their research activities. The importance of resources and equipment for making discoveries or innovations was best captured by Stephan and Levin: "In the world of science, the bird who gets the worm is not necessarily the brightest bird or the hardest-working bird, but the bird with the best equipment for worm spotting and worm digging" (1992:13).

In spite of official and unofficial restrictions on employing women and faculty wives at universities, institutional backing gives women who want to do significant work in science the resources and support (although inadequate) to pursue and continue their work. It would be very difficult for any female or male scientists to "stay in the game" without some kind of formal support from universities or research institutes. Even someone as independent and determined as Barbara McClintock needed continuous external support for her pioneering research on maize. She was pleased that, as a Carnegie Institute employee, she did not have to apply for federal grants. She finally made Cold Spring Harbor, a research and conference center, her professional home. Comfort contended that Cold Spring Harbor was a factor in the McClintock renaissance (2001:254). Institutional support was also vital to others in this study. Marie Curie taught physics at the Normal School for Girls in Paris. The Radium Institute became the professional base for Marie Curie and Irene Joliot-Curie. The American Museum of Natural History was Margaret Mead's professional home for half a century. In addition to having a permanent base at the museum, she served as chair of the department of social science and professor of anthropology at Fordham. Mead was also an adjunct professor at Columbia while holding several visiting lectureships at different times.

However, financial support was not the bulk of this support. Most academic women were underpaid at the time. Maria Goeppert-Mayer had a paid job as a senior physicist at Argonne National Laboratory outside Chicago. Before landing a job at the University of Chicago's Fermi Institute and later at the University of California in La Jolla, Goeppert-Mayer was a volunteer researcher or lecturer at two other leading research universities—Johns Hopkins and Columbia. Although Rachel Carson had no formal academic affiliation, she drew extensively from her ties established within and beyond the U.S. Bureau of Fisheries for her environmental research. Washington University in St. Louis, Missouri was Rita Levi-Montalcini's professional home for almost three decades. Dorothy Hodgkin did her postdoctoral work at Cambridge under the supervision of John Desmond Bernal. Her whole career was built on her position at Somerville College, a women's college. The Bronx VA Hospital was Rosalyn Yalow's professional home for three decades. She was there as a researcher and administrator. The University of Pennsylvania has been Fay Ajzenberg-Selove's niche for decades after she left her tenured position at Haverford College.

Another function of having institutional affiliation is that it confers some sort of formal status on women as "scientists" in the professional community. These female pioneers seldom occupied the same rank or title as their male colleagues or scientific spouses. However, the symbolic titles of "lecturers," "research associates," or "research assistants" allowed these women to pursue their work in a stimulating scientific environment. Even within the confine of these settings, as well as facing the prospects of title deflation, these female scientists could engage in meaningful intellectual exchanges with their colleagues in and outside their immediate environment. Through publication as well as participation in conferences and other professional activities, they gradually built up their reputation in the scientific establishment. Additionally, maintaining ties with academic or research institutions offered them the opportunity to "stay in the loop" by being a part of the invisible college, or by forming their own networks. In short, institutional support provided both symbolic and practical value to the women in this male domain.

## SCIENCE

The rise of professionalism in science in the late nineteenth and early twentieth centuries presented both challenges and opportunities to female scientists. By then, women already had a growing presence in professional scientific societies. To improve the status and prestige of science, many of these national societies raised their standards for membership. The new requirements entailed different qualifications for male and female scientists. Now, rather than officially barring women from entry, many national societies required women to possess higher qualifications than men to gain admission (Pycior et al. 1996:143, 230–231). As a result, male scientists continued to maintain higher professional standing compared to female scientists in this two-tiered system. Rossiter called this standard practice of admitting women but relegating them to a lower or subordinate level a "scientific society dilemma" (1982:73).

Though far from withdrawing their participation in mainstream professional organizations, the professionalization of science compelled female scientists to take actions to improve their marginal status in the scientific community. In response to systematic attempts to restrict women's participation in predominantly male scientific associations, female scientists formed their own local women's clubs in science. For the most part, these women's science clubs served as quasi-professional scientific societies conferring honors, awards, and fellowships to female students and prominent female scientists.

It is difficult to gauge the overall impact of these subcultures on women's advancement in science. However, these clubs, though far from challenging mainstream scientific organizations, provided a platform for women to expand the new/old-girl networks. Historical evidence suggests that this was indeed the case. As a group, female scientists had taken constructive steps to minimize the negative effect of professionalization on their already-marginal status in the scientific establishment. Nonetheless, some scholars like Rossiter (1982) had reservations about the effectiveness of these women's clubs. Critics noted that these auxiliaries, which are primarily social in nature, have done little for women, but instead enhanced women's segregated status in the scientific community.

From the experiences of female scientific pioneers, there is evidence that the new/old girl networks were seldom used by female scientific pioneers for career development or advancement. Instead, subjects in this study relied on limited individual ties with women to seek career advancement: (a) Irene Joliot-Curie's ties with Marie Curie; (b) Margaret Mead's lifelong, intimate relationship with Ruth Benedict; (c) Barbara McClintock's close association with Harriet Creighton; (d) Rachel Carson's loyalty to Mary Skinker, Marie Rodell and her administrative assistant; and (e) Rosalyn Yalow's acquaintance with Edith Quimby. One exception was Fay Ajzenberg-Selove, who received enthusiastic support from female colleagues when she disputed her tenure case with her employer.

There are three explanations for a lack of female pioneers in this segment of the invisible college: (a) an underdevelopment of these women's networks, (b) their perceived weaknesses in enhancing women's progress in science, and (c) the availability of alternative means for maintaining ties with the scientific establishment.

The rise of professionalism in science has presented another dilemma for women. To gain acceptance of their work in the scientific community, even female scientific pioneers have to downplay their gender in public communication. Rachel Carson is one example. In her correspondence with Edward Weeks, editor of *Atlantic Monthly*, regarding her submission, she used "R.L. Carson" in her signature. The reason for not using her full name was to take advantage of the popular image of scientists as men:

> From time to time in my work with the Bureau of Fisheries . . . I am called upon to prepare articles on commercial fisheries. In as much as these articles deal largely with economic questions, the scientific basis of conservation measures, etc., we have felt that *they would be more effective . . . if they were presumably written by a man*. For various reasons I prefer to use the same name in my personal writing [emphasis added] (Lear 1998:87).

Greenstein admitted that one needs to be aggressive and competitive to win at ball games. But he questions the necessity of being competitive in order to be successful in science, if a person simply wants to "unlock the secrets of nature" (Greenstein 1998:196). The rise of professionalism in science has raised the stakes for practitioners who attempt to seek opportunities, resources, and recognition. Preceding discussions highlight the reciprocal relations of science and society and, more importantly, the aggressive and competitive styles among female scientific pioneers.

Female scientific pioneers profiled in this book managed to use the institution of science to their advantage. All of them were able to establish and maintain strong ties with influential male scientists. These exceptional women were able to do what many female scientists have failed to do: create important alliances in science. Even someone as reclusive as Marie Curie had a great deal of support. Most of her supporters were American men and Eastern European women. Through their connection with a group of leading male scientists, whether scientific spouses, mentors, or sponsors, female scientific pioneers were in a complex web of information, resources, and opportunities. Most important to their careers, these influential male scientists became powerful allies and staunch supporters. Why would these powerful male scientists do that for their female colleagues? Rossiter (1995:331–332) observed that they gained support and sympathy from their male colleagues because these women were so hardworking and dedicated to their work, and that their work had broken new grounds. There is some support for this assertion in male-oriented professions (Driscoll and Goldberg 1993; Epstein 1970). For example, anthropology was still a male-dominated field at least until the 1970s. In her early 20s, Margaret Mead worked directly with Franz Boas and gained entry into his inner circle. Impressed by Mead's energy, intellect, and enthusiasm, Boas, along with scholars such as linguist Edward Samir and cultural anthropologist Ruth Benedict, welcomed her company.

Other female scientific pioneers in this study had similar experiences. Organizations and gatekeepers of science saw these women as resources, assets rather than threats. One can argue that large institutions and those with significant power are on the lookout for great ideas, so they are not afraid to invite "newcomers" (including women) to join them and work on their projects.

## IMPLICATIONS

A host of factors is responsible for the successful pursuit of pioneering scientific work by women. Aside from individual talents and effort, sociologists of science have underscored the importance of being brought up and trained in stimulating environments for the advancement of successful female scientists. This chapter examined three dimensions of institutional influences on career progress: family, schools/institutes, and science. These forces combine to elicit career success in science.

Family is the institution that has the most enduring impact on career aspirations and career development of female scientific pioneers. Family involvement in education during childhood is essential for achieving success in science. There is some indication that female scientific pioneers identify strongly with the male figures in family, usually the father. Also, Rossi found close father-daughter relationships among famous female mathematicians (Dash 1973:257). However, (grand)mothers played a major role in the early socialization of female scientific pioneers as well. Generally, parents are a decisive factor in childhood and adolescence. Yet as we have seen, their direct influence on personal and professional development diminishes in adulthood, and schools, institutes, peers, and science increase their influence on the career advancement of pioneers-to-be. Support from academic or research institutions is critical in their career development—as a training ground, a major source of research support, and a gateway for networks.

The rise of professionalism in science in the late nineteenth and twentieth centuries presented both challenges and opportunities to female scientists. A close examination of their life and career histories reveals that these talented, hardworking women were able to develop new ideas and make significant discoveries because they happened to be in a place that was intellectually stimulating and resourceful as well as a time that was ripe for scientific breakthroughs (Kuhn 1962; Stephan and Levin 1992). One can say that many of these women were indeed *at the right place at the right time.*

Indeed, institutions have played an increasingly important role in changing women's prospects and progress in science. The U.S. National Science Foundation, for example, commissioned a series of initiatives to make universities a better place for women in training, teaching, and research. However, there may be limits to what social engineering can do to improve the advancement of women in science. A billion-dollar program, Canada Research Chair (CRC), launched by the Canadian federal government recently to improve world-class research in universities, has been criticized for its gender bias. Scientists appointed by the universities as CRC receive a substantial amount of funding and support for their research. However, due to its emphasis on hard science, women who are underrepresented in these fields are seldom nominated by their university for these highly coveted appointments (*Chronicle of Higher Education* 2004).

Women with outstanding careers in science *happened* to be in the right place, yet perhaps their experience was more deliberate than that. As discussed in the forthcoming chapter, many of them received shrewd advice that helped them avoid the pitfalls others emcountered.

*Chapter Six*

# The Contradictions of Norms

## INTRODUCTION

This chapter revisits the central ideas and main arguments in the field of sociology of science. The question central to this chapter is: *What do the experiences of female scientific pioneers tell us about the normative structure of science?*

Norms are the ideal institutional standards for ensuring "the rationality and the basic character of scientific knowledge" (Mitroff 1974:11). For example, the normative structure of science dictates that careers are open to talent and rewards go to those with excellent performance (Zuckerman 1988:518–520). Counternorms are departures from standard policies and practices in the scientific establishment. Were female scientific pioneers following different strategies from their male counterparts to advance their career goals?

The following discussions focus on three issues:

1. Is science a meritocractic institution?
2. What are the most critical elements for achieving success in science?
3. Is there a typical linear career path to scientific success?

It is true that most of the formal barriers women faced in science have now been dismantled. However, many female scientific pioneers in this study were confronted with both formal and informal obstacles throughout their careers. By addressing these questions in light of their experiences, we can probably see these underlying forces more clearly.

## PREFERENTIALISM AND SELF-INTERESTEDNESS

There is mixed support for the ideology of meritocracy in science. For female scientific pioneers, science is not an entirely meritocractic institution. What is most intriguing is that the career experiences of subjects in this study are highly consistent with the results of studies on male and minority scientists—that the scientific reward structure is neither wholly universalistic nor particularistic (Zuckerman 1988:518).

According to universalism, one of the four ethos of science, the evaluation of an individual's ability and performance should be governed by a set of objective, impersonal standards. Gender, race, and other functionally irrelevant characteristics should have no bearing on the assessment of a person's work. Disinterestedness, another norm of science, stipulates that scientists act in a selfless manner in advancing scientific knowledge (Merton 1973; Zuckerman 1988:515). Both philosophies mean that scientists are expected to withhold personal bias in evaluation. As stated in the opening of this book, "puzzlement" (curiosity) is the primary reason for doing science. The "gold quest" (financial rewards) and "blue ribbon" (recognition) are secondary concerns. Clearly, none of the subjects in this study would have been able to attain what they did if their work was not truly outstanding in the view of their colleagues and contemporaries. Therefore, we can rule out a lack of talents or insufficient effort as an explanation for the educational and occupational barriers they encountered. Proponents of the claim of universalism could use the exceptional achievements of female Nobel laureates in science to bolster their assertion that the meritocractic system works in science. As shown below, this is a rather narrow view of the scientific community many female scientific pioneers have participated in.

On the other hand, one can make the case that performance in science is often governed by subjective judging. Because science is a social activity, fair evaluations are generally hard to come by, and remarks given are seldom totally objective. Studies have also documented the limitations of using citations to measure the quality or importance of scientific work. Skeptics have forcefully argued that citation counts are, at best, crude indicators of *perceived* quality of work. Equally important, citations list only the first-named authors in multiple-authored scientific papers (Inhaler and Przednowek 1976:35).

Scholars from different fields have noted the prevalence of particularistic standards in policies and practices; they have argued that the ideology of meritocracy does not always work in science (Cole 1992; Comfort 2001:255; Mitroff 1974). Ernest Rutherford along with others, for example, questioned the originality of Marie Curie's research, and attributed her scientific success to hard work and tenacity rather than creativity. When she received the second Nobel Prize, physicist Bertram Borden Boltwood made similar remarks, stating, "Mme. Curie is just what I have always thought she was, a plain darn fool, and you will find it out for certain before long" (Ogilvie and Harvey 2000:314–315).

As noted earlier, not everyone who has done outstanding work is rewarded or recognized by the scientific community. To challenge the claim of universalism, critics could cite numerous examples of female inventors or innovators who have been underrecognized or underrewarded for their achievements or contributions to science (Creese 1998; Kohlstedt 1999; Mozans 1991; Stanley 1995). Some have investigated the implications of preferentialism and self-interestedness for women in science. The "Matthew effect" depicts the cumulative advantage that men have over women in science (Merton 1973): Men's relatively early entry into the scientific professions gives them an edge over women—the newcomers—when competing for opportunities, resources, and recognition. Even very small differences in performance matter at the early stages of competition. This initial edge can be translated into enormous differences in career outcomes. Without any intervention, this cumulative advantage can contribute to a widening gap in career achievements between those who have this advantage and those who do not.

Pierre Curie was a beneficiary of the Matthew effect in science (Pycior 1993:316). Compared to his scientific wife, Marie Curie, Pierre Curie continuously received more resources and recognition from the scientific community for career advancement. Williams (1995) made a similar observation of men doing women's work. Men in traditionally female occupations such as nursing, social work, elementary school teaching, and librarianship tended to receive more support than their female counterparts for advancement to leadership positions. These male workers experienced what Williams called the "glass escalator effect" because of their speedy ascension to the upper echelon of organizational hierarchy. Williams attributed men's hidden advantages in female occupations to a "spillover" of traditional gender-role expectations from the society to the workplace—that men are generally expected to be leaders.

There are other situations where men tend to fare better professionally than women in science, all things being equal. The "Matilda effect" in science postulates that a man is more likely to receive the bulk of credit for collaborative work with a woman (Rossiter 1993). At the early stages of their career, Pierre Curie's name tended to be more visible than Marie Curie's, even though both were listed as coauthors on their publications. To offset the Matilda effect, Marie Curie adopted a different publication strategy. To counteract the stereotypes of women as subordinates in scientific publications, Marie Curie learned to claim proper credit for her individual work as well as collaborative work with Pierre Curie and other male scientists (Pycior 1993:302).

Cumulative advantage undercuts intergender competition in science, preserves male domination in science, and maintains stratification in science. What are the implications of the Matthew and Matilda effects for male and female scientists? How useful are these two concepts in understanding the careers of scientific pioneers? Career experiences and outcomes of subjects in this study suggest that the operation of Matthew and Matilda effects in science certainly has an adverse (but not devastating) impact on women's career progress. In this chapter, I attempt to demonstrate that female scientific pioneers have turned the semimeritocractic structure in science to their advantage.

Subjects in this study, along with other female scientists, have experienced employers' or editors' taste for discrimination; there had been a trend of replacing women with men at women's colleges. The rationale for preferring hiring men to women was to enhance the reputation of women's colleges, and these hiring practices were not restricted in women's colleges. Employers' taste for discrimination was also found in male-dominated and coeducational institutions. As a result, many female scientists with Ph.D.s were unable to find employment at major academic institutions (Lear 1998:41; Rossiter 1982, 1995). Their predicament lends credence to the claim that a link between individual and institutional competition can be found in scientists' careers (Hagstrom 1974:16). Individual competition can result in career shifts only if institutions are flexible enough to offer new positions. A review of the career trajectories of the sample clearly shows that their career ambitions were repeatedly stalled because of legal and other structural barriers.

Even employed, many female faculty were underutilized. Most obtained what Fox and Stephan call employment "left-overs" (2001:119). Mathematicians Emmy Noether (1882–1935) and Grace Chisholm Young (1868–1944) as well as sociologist Helen Hughes are perfect examples of underutilization among female scientists (Hoecker-Drysdale 1996:225–231; Wiegand 1996:139). Emmy Noether, with a Ph.D. from Gottingen University in Germany and despite the intervention of her male peers, could not even obtain a position of junior lecturer in mathematics (1908–1915). Grace C. Young, with a Ph.D. from the same university, did not receive equal credits for joint mathematical work with husband Will H. Young (1898–1913). Helen Hughes, with a Ph.D. from the University of Chicago and wife of prominent sociologist Everett Hughes, worked 17 underpaid years as an editorial assistant and later managing editor of the *American Journal of Sociology* (1944–1961).

There is evidence of a continuation of these practices in recent decades. In a study of gender differences in advancement to higher academic rank, Long,

Allison, and McGinnis (1993) found that women tend to hold "courtesy appointments" at universities. Compared to men, they are less likely to be promoted to associate and full professor ranks. This is, in part, because women are expected to meet higher standards for promotion. More importantly, there are more obstacles for women working in more prestigious departments in terms of promotions than those in less prestigious ones.

Female pioneers in this study also had extensive experiences of being underutilized, underrecognized, or underrewarded. These instances ranged from having positions incommensurate with their qualifications, to failure to gain admission to prestigious scientific societies, to not being considered for nomination. Talented and determined female scientists who have male collaborators are affected by subtler forms of discrimination. One example is "scientific backbiting." As mentioned earlier, Marie Curie had repeatedly been second-guessed by the scientific establishment for her joint work with Pierre Curie. Many scientific elites did not consider her as her husband's equal.

Maria Goeppert-Mayer is another example. She thoroughly investigated the nuclear shell structure, yet her scientific reputation was constantly questioned. Many attributed her acknowledgment at the end of the scientific paper "Closed Shells in Nuclei II" to Fermi's responsibility for her winning of the Nobel Prize: "Thanks are due to Enrico Fermi for the remark 'Is there any indication of spin orbit coupling?' which was the origin of this paper" (Jones 1990:203). Goeppert-Mayer coauthored *Statistical Mechanics* with her husband Joseph Mayer. Others considered her merely as her husband's editorial assistant (Dash 1973:289).

Similar to Marie Curie's scientific spouse, Pierre Curie, Rosalyn Yalow's scientific partner, Solomon Berson, was often considered the creative part of the research team.

Rita Levi-Montalcini had similar experiences of being ignored for the discovery of the nerve growth factor (NGF):

> For so many years there was no doubt that I was—as I really was—responsible for the discovery. For a long time people didn't mention how NGF was discovered. My name was entirely left out of the literature. People repeated my experiment and didn't mention my name! I am not a person to be bitter, but it was astonishing to find it completely canceled (Liversidge 1988:72).

Scientific backbiting has also been directed at male team members. Experiences of the male partners of two female scientific pioneers are telling. After teaming up with his famous scientific spouse, Irene Joliot-Curie, Frederic Joliot worked hard to establish his professional reputation. In the early stages of his career, he was often criticized for using the Curie name to enhance his position in the scientific establishment. Hamburger invited Levi-Montalcini to St. Louis and became her first scientific partner. It was Stanley Cohen, however, not Viktor Hamburger, who shared the Nobel Prize with her for their discovery of NGF. While many in the scientific establishment thought highly of Hamburger's contribution to its discovery, Levi-Montalcini vehemently denied his pivotal role in the study of NGF:

> Viktor Hamburger was not there when I made the discovery of the soluble agent released by NGF. He had no participation in this. He was in Boston, I was in Rio de Janeiro all by myself when I discovered how to elucidate the way NGF works. So I believe I really am the discoverer of NGF (Liversidge 1988:72).

This dispute over priority might be owed to the fact that Levi-Montalcini accepted Hamburger's invitation to do postdoctoral work in Washington University, or to their longtime research collaboration. These events led to the public's perception of Hamburger as a codiscoverer of NGF.

Overall, the experience of female pioneers in science tells us that the notions of universalism and disinterestedness cannot fully capture the complexity of their career progress. Apparently, both universalistic/particularistic standards and disinterestedness/self-interestedness were in operation throughout their careers. Once they overcame the initial barriers for women to obtain science training, other functionally irrelevant characteristics (such as gender, nationality, and religion) became more salient as they managed to advance their careers and were competing for resources, fame, and recognition with their male counterparts.

Subjects in this study did not enjoy prospects for education and employment comparable to those of men at the time. The most straightforward evidence for the existence of preferentialism and self-interestedness in science were the formal restrictions on women in graduate schools. Historically, women were not formally or officially admitted into graduate schools, regardless of their qualifications. Only a few women who were interested in pursuing advanced studies were given the status of "special student" to receive informal university training (Rossiter 1982:31). If its opportunity structure were primarily merit-based, more women would have been able to receive advanced training in science.

Setting higher admissions standards into national scientific societies for women is another piece of evidence that evaluation of the qualifications of female scientific pioneers and other female scientists is not subject to the same requirements used for assessing the qualifications of their male peers. As a result of the creation of separate levels of membership, a relatively large proportion of male scientists were in the higher rank of "fellows," while female scientists, even if they met the more stringent admissions standards, were concentrated in the lower rank of "members" (Rossiter 1982, 1995). Due to women's lower ranking in major scientific societies, women were less likely than men were to sit on advisory boards or nomination committees for honors and awards. Put simply, women are less inclined than men to be gate-keepers or rainmakers in science. One can then argue that a person's prominence in science or potential impact on the scientific reward structure is not primarily based on the excellence of his work.

The implication for women who do pioneering work is that their ideas and discoveries are less likely to receive proper recognition from the scientific establishment. The fact that Pierre Curie, but not Marie Curie, was initially considered for nomination (with Henri Becquerel) for the Nobel Prize in physics reveals inherent weaknesses in the scientific honor system and challenges the normative structure of science. Another counterexample of the norms of science is that, despite their extraordinary accomplishments, Marie Curie and Irene Joliot-Curie were both refused admission to the French Academy of Sciences.

Many female scientific pioneers were keenly aware of preferentialism and self-interestedness in evaluation and allocation of rewards. To avoid gender bias by editors of scientific publications (usually men), some female authors adopted the strategy of hiding their gender. For example, like Rachel Carson, they used their initials in professional correspondence (Lear 1998:87; Pycior 1996:130; Rossiter 1982:262). Of course, this practice ultimately lost its effectiveness when their work began to attract attention from the scientific community. Unless one publishes under a pseudonym, it is very difficult to hide the actual identity of an author permanently. Mary Ann Evans, a British author and scholar, published under the name George Eliot (Howe 1999:207). Reverend John A. Zahm published *Woman in Science* under the name H.J. Mozans (Mozans 1991). This is just another passive tactic to accommodate and reinforce rather than challenge existing gender bias. The outcome is that the patriarchal relationship in scientific communities remains intact.

The prevalence of antinepotism rules from the 1930s to the 1960s had a dampening impact on the career prospects of female scientists. As noted in Chapter 4, Maria Goeppert-Mayer was a longtime bearer of deflated titles at various institutions while researching the nuclear shell structure. Goeppert-Mayer's experience was neither unique nor unprecedented. Another example of antinepotism was that, after winning the Nobel Prize in physics with Pierre Curie, Marie Curie received an appointment as the director of Pierre Curie's research lab instead of a professorship, as he had. Taken together, these policies and practices support the claim of the existence of "job queue" and "gender queue" in the scientific community. According to Reskin and Roos

(1990), all other things being equal, employers tend to prefer male workers to female workers, and desirable jobs are generally reserved for men. One can make the argument that there is evidence for the employer's inclination to discriminate even in fields that profess to operate on universalistic standards and disinterestedness.

If title deflation is detrimental to women's career advancement in science, the outstanding accomplishments of female pioneers such as Marie Curie and Maria Goeppert-Mayer are counterintuitive. How can we reconcile the fact that these female scientists held a marginal academic or research position but still managed to get ahead in science? To understand why they became pioneers in their fields, it is necessary to look beyond the formal mechanisms from which these female scientists drew their support. I argue that they were able to offset the adverse effects of title deflation on their career advancement through what I label "compensatory mechanisms." Almost without exception, their scientific spouses or scientific colleagues (predominantly male) became the most important asset for them. The direct and indirect support from a few influential male figures in science compensated their loss in professional status and occupational rewards in the scientific community. In addition to being a sounding board for new ideas, the scientific spouses or sponsors literally assumed the unofficial role of "agents" for these female scientists. These agents, in one way or another, always looked out and defended the professional interests of their "clients." Equally important, both viewed each other as an invaluable ally rather than a potential rival in the scientific community.

For the most part, the phenomenon of title deflation did not define or restrict the role of these women in the scientific establishment. Their accomplishments suggest they succeeded in growing out of the subordinate role of women in science, as well as in creating a niche for themselves in the scientific community. The following observation by McCay, in her biography of Rachel Carson, best captures the challenges subjects in this study faced in advancing their careers:

> If a woman did not have a powerful mentor, she could not hope to get any position. Even if she did have a male on the faculty who supported her research and gave her assistantships and lectureships, her chances of becoming anything more than an assistant professor were slim. The prejudice against women in academic science departments was so intense that even the most brilliant female protégés of prestigious male scientists were often passed over for promotion despite their mentor's support (McCay 1993:11).

This means that succeeding in science requires more than simply having powerful allies and winning the support of influential backers. What *must* a woman do to become a pioneer in science?

## NETVIGATORS AND HETEROSOCIAL REPRODUCTION

Another interesting finding in the study was that networking matters. Results corroborate the claim that lineage and network are critical determinants of great accomplishments (Zuckerman 1996). Subjects in this study were *netvigators* who actively pursued and developed social and professional networks. All were engaged in strategic practices typically used by male scientists for career advancement. Through regular correspondence with a selected group of renowned male scientists, these female scientists made their new discoveries, ideas, or work known in the invisible college. This routine of prepublication opened up possibilities for collaboration as well as nominations for scientific honors. Aside from being a source of support and encouragement, male supporters often alerted them of new career and funding opportunities. Without exception, all subjects in this study are perfect examples of how to use this tactic to gain entry into the *old-boys club*. All this suggests that being a netvigator is essential for aspiring female scientists to gain notoriety for their work and pave their way to join the ranks of scientific pioneers.

Despite striking differences in personality and work style, subjects in this study had three commonalities. They (a) studied under or worked with great masters in the field, (b) had an apprenticeship capacity, and (c) were interested in timely strategic engagement. All of these are crucial elements in the success of male scientists, too (Booth 1989; Clark 1982; Cole 1979; Long and Fox 1995; Rose 1989; Zuckerman 1996). Many eminent scientists, including female pioneers in this study, acknowledged the profound influence of their teachers on personal and professional development (Wolpert and Richards 1988).

Being plugged into the old-boy network has been a crucial ingredient of career success for scientific elites. This approach is simple in principle but difficult in practice. A careful review of the career histories of the sample revealed that they managed to overcome the formal and informal obstacles in the scientific establishment and became a part of the inner circle (Zuckerman, Cole, and Bruer 1991). To get the credit they deserved for their work, they understood that "who they knew" was as important as "what they knew." Building on the homosocial theory of sex roles (Lipman-Blumen 1976), I argue that these female scientific pioneers managed to tap into the male homosocial world in science. (A female homosocial world in science has not yet evolved.) For the most part, the male scientists they associated with were people with authority and status inside and outside science. These people ran the departments or institutions and wielded (in)formal power within the scientific establishment. Female pioneers-to-be were able to build their scientific reputation and gain indirect access to resources through relationships with scientific spouses, teachers, mentors, sponsors, or colleagues.

So how did these women become, in the eyes of their male colleagues, one of the boys in the invisible college (Crane 1972)? Very early on, these talented women actively managed their own careers and continued to be proactive in building their careers. They all had the ability to build important, long-lasting ties. In simple terms, they had their ways of finding the person(s) who knew where the support or resources could be found. These people became their "scientific partners" who helped them overcome difficulties and, at times, radically changed the course of their research and careers. They succeeded in developing collegial ties that have characterized relationships among male scientists (Schiebinger 1987:322). In short, these women were subtly engaged in self-promotion. I call these tactics "timely strategic engagement."

Throughout their lives, subjects in this study had cultivated loyalty and trust with a small group of influential or powerful figures—their professors, spouses, or colleagues, whom I label as "agents." By many accounts, these agents could be considered "secure leaders" in science; they were not afraid to surround themselves with (or to hire) people, men or women, who were better (and younger) than they were. These agents showed genuine concern for their client's (the female scientific pioneer's) progress in education and work, and did a number of things that were absolutely necessary for their client's career development and success, including: (a) raising the bar on performance, (b) sparking changes in research direction, and (c) encouraging experimentation and innovative research. On the other hand, the clients understood the importance of keeping close contact with the most precious ally in their own career development and career advancement. Through written correspondence and other types of professional activities, these female scientists regularly communicated their enthusiasm and devotion to scientific work to their agents. They also used their agents as gateways to other potential alliances in the scientific community, such as advisory board members or nomination committees. In other words, their agents performed the vital functions of "satellite" and "lobbyist." They became literally a bridge between female pioneers-to-be and the scientific establishment.

More importantly, these agents were willing to go to any lengths to help their clients when necessary. There are numerous examples that their agents lobbied for their work or career plans, with or without their knowledge. Pierre Curie helped Marie obtain lab space for her work and became her interface with the French scientific community. When he learned from the Swedish mathematician Gosta Mittag-Leffler that she was not nominated for the Nobel Prize, he let Mittag-Leffler know that he wished to be considered with Marie Curie for the honor (Pycior 1993:317). (Mittag-Leffler had supported Sofia Kovalevskaia's successful candidacy for a mathematical professorship at Stockholm University.) As a result of Pierre Curie's intervention, this scientific couple jointly received the prize in 1903. Further, the egalitarian scientific partnership with Pierre Curie allowed Marie Curie to gradually establish her scientific reputation in radioactivity. She was able to claim credits both for their joint work and for her own work. Although Marie Curie shunned publicity as well as politicking in France, her successful fundraising tours to the United States orchestrated by an American journalist, Marie Meloney, revealed her tact. Her appeal stemmed from her roles of being a female scientist and a mother, as well as the hope of finding a cure for cancer. Later in her life, in addition to running her lab, Curie became more active in public and professional communities. She participated in war efforts, as noted earlier, and was also involved in the League of Nations' International Committee on Intellectual Co-operation. She never stopped maintaining friendships and correspondence with leading figures in the scientific community, including Albert Einstein (Ogilvie and Harvey 2000:315).

Because of her parents' friendships with renowned scientists, Irene Joliot-Curie was born into the inner circle. As a child, she met many leading scientists at home and in outdoor family activities, including Einstein. Her high school teachers were key scientific figures in their fields, and she socialized with their children. For Joliot-Curie, networking was a way of life. She always expected and received support from her parents' friends and colleagues. Until her nominations for the French Academy of Sciences were turned down, she had never been excluded from the scientific establishment. Despite her privileged background, Joliot-Curie realized the importance of forming crucial alliances. Throughout her career, she had two very successful scientific partnerships. Compared to other female scientific pioneers, her scientific collaborations—maternal/spousal—were one of a kind. In the first collaboration, which lasted for decades, Joliot-Curie was the apprentice of Marie Curie, her mother as well as a two-time Nobel Prize winner. A role reversal took place in the second, which lasted for about five years: She was the master of a novice scientist, her future spouse and later a cowinner of the Nobel Prize in chemistry in 1935—Frederic Joliot.

Fate and luck dictated her choice of collaborators. After the tragic death of her father, Joliot-Curie became the natural scientific partner for her mother. Marie Curie became her senior partner. Joliot-Curie learned math from Marie Curie in the co-op school set up by her mother and colleagues. She worked first, side by side with her mother, and later, independently at the war front. At the Radium Institute, she was her mother's research assistant. Her choice of polonium as a dissertation topic was blessed by Marie Curie, who had control over its supply in France. Slowly and surely, Joliot-Curie became Marie Curie's successor. Joliot-Curie's grooming by Marie Curie constitutes a classic example of homosocial reproduction. After the loss of her father, no one was in a better position than Joliot-Curie to be Marie Curie's collaborator. Curie's seclusion helped bring Joliot-Curie closer to her mother. Her mother even picked a junior collaborator for her. When Fred Joliot arrived at the Radium Institute, Marie assigned this newcomer to work with her daughter. Fred, a novice scientist, and Joliot-Curie later formed a successful and powerful spousal research team in France.

Would Joliot-Curie's career trajectory have been any different under the tutelage of both Marie and Pierre Curie (i.e., if her father had survived the tragic accident)? Likely, Joliot-Curie would have been very successful even if she missed the opportunity to collaborate closely with her mother.

If networking is an art, Margaret Mead perfected it. An examination of her extended list of professional activities suggests that giving and getting favors was second nature for Mead. Her enthusiasm, along with her outgoing and outspoken style, allowed her to gain a foothold in the scientific establishment. As a graduate student, Mead understood the significance of networking on reputation building. She studied under the renowned German-American anthropologist Franz Boas, who was then the head of Barnard College's anthropology department. An expert in the field of anthropology, Boas had made enormous contributions in linguistics, archaeology, and physical anthropology. Not long after, Mead became a part of the inner circle of her teacher, mentor, and sponsor, Franz Boas.

Mead's inclusive (collaborative) approach to research and scholarship reflected her willingness to break out of the traditional mode of doing anthropology. Mead's prolific scholarship could be attributed in part to her ability to continuously seek out people with complementary skills and to engage them in collaborative work. Her collaborators included graduate students and research assistants as well as scholars from other disciplines. Additionally, Mead became an icon of "scientific couples" in the United States. Over her entire career, she formed notorious scientific partnerships outside the United States with two of her three husbands—Fortune and Bateson—both of whom were non-American-born social scientists. With Reo Fortune, a New Zealander, she conducted fieldwork in New Guinea. Mead, along with her British third husband, Gregory Bateson, conducted fieldwork in Bali and New Guinea (Mark 1999). None of this diminished Ruth Benedict's intellectual influence on Mead because of their less well-publicized but intimate relationship (Banner 2003; Lapsley 1999). Nevertheless, Mead's risk-taking and innovative style paid off in light of her professional accomplishments and enduring influence in and outside the scientific community.

When it came to networking and sponsorship, Mead was the opposite of the mother-daughter team (Curie and Joliot-Curie). Marie Curie shunned publicity and closely guarded her privacy. Irene Joliot-Curie took her membership in the inner circle for granted. For the most part, Joliot-Curie confined her activities to the Radium Institute. In contrast, Mead sought and embraced attention from both the public and professional communities.

Of all the subjects in this study, Mead was the one who truly enjoyed her celebrity status. Further, Mead consistently reached out to people from different fields and different sectors, and collaborated with people from diverse backgrounds. However, the Boas-Mead team was the counterexample of homosocial reproduction. Mead began to work with Boas in her early 20s. His achievements and ideas had a strong imprint on Mead. In light of her pioneering work, career trajectory, and influence in cultural anthropology, Mead followed in his footsteps.

Mead's network-building strategies were inclusive; she brought together people with complementary skills through marriages or long-time collaborations. Barbara McClintock was the direct opposite of Mead in many respects. She was never married and liked to work alone, independently, and freely. Within the scientific establishment, Mead was a natural-born empire builder, while McClintock was a sojourner. However, because science is a social activity, it is unlikely that McClintock could have managed her way to the top of science without drawing on the support of others. To enter and participate in the inner circle, Mead cast a wide net to create both sparse and expansive ties through marriages, friendships, and collaborations. McClintock was selective in cultivating formal and informal ties. Relying on lifelong friendships with a few male colleagues, McClintock was able to tap into the scientific community's resources if and when necessary. During her entire career, she drew intellectual and emotional support primarily from two geneticists. Rollins A. Emerson, a renowned maize geneticist and head of the plant-breeding department, brought Marcus Rhoades and George Beadle to Cornell. These two talented graduate students found McClintock's cytological work exciting. She was closely associated with Rhoades and Beadle. Together, they worked on the questions generated by McClintock's cytological work. She reported this close-knit professional cross-sex relationship as mutually beneficial:

> We were opening up a new field, the three of us were. . . . We were a group, all of us highly motivated, and we used to have our own seminars from which we'd exclude the professor—just us and a few others (Keller 1983:48–51).

What drew these two young male scientists to her? It was McClintock's brilliance and uniqueness. As Rhoades put it:

> I recognized from the start that she was good, that she was much better than I was, and I didn't resent it at all. I gave her full credit for it. Because—hell—it

was so damn obvious: she was something special. . . . I've known a lot of famous scientists. But the only one I thought really was a genius was McClintock
(Keller 1983:50).


Beadle concurred. "She wasn't like the rest of us. . . . She was so good"
(Keller 1983:51). In his opinion, McClintock's work was "fantastic," "spec-
tacular" and "the best job that's been done."

This trio benefited professionally from this early kinship; each went on to
have distinguished careers in science. Rhoades became a leading geneticist
and Beadle shared a Nobel Prize in physiology in 1958.

Lewis Stadler, a colleague of Emerson's at Cornell, was another champion for McClintock. Stadler finally found McClintock an assistant professor position at the University of Missouri. Despite having difficulties in
finding a suitable teaching position, she continued to receive tremendous
support from rainmakers at critical stages of her career. Support from
Carnegie President Vannevar Bush, for example, solidified her research position at Cold Spring Harbor. Stadler and Emerson were members of the National Academy of Sciences when McClintock became the third woman in
that prestigious society's history to be an elected member. However, Mc-
Clintock did not rely just on her male friends, who had regular contacts, but
also on Harriet Creighton, a fellow graduate student and lifelong friend with
whom she consulted extensively.

Like Irene Joliot-Curie, Maria Goeppert-Mayer was born into a web of professional ties. Living in a university town, Maria grew up seeing and interacting with intellectuals and their offspring. Her father was her strongest supporter from her childhood to young adulthood. Friedrich Goeppert's
expectation of his only child to be more than "just a woman" set the direction
of Goeppert-Mayer's academic and professional development. She also
stayed *extremely* well-connected throughout her career. One could argue that
her lifelong supporters were among the who's who in physics (McGrayne
1998:184).

Max Born was her mentor when Goeppert-Mayer was studying for the doctorate. After her father died, he became a father figure to her. James Franck
also had a very high regard for Goeppert-Mayer. Her doctoral examining
committee consisted of three Nobel laureates. James Franck received the
1925 Nobel Prize for physics, Adolf Windaus was awarded the 1928 Nobel
Prize for chemistry, and Max Born received the 1954 Nobel Prize for physics.
She maintained social and professional ties with these influential male scientists and their families after her relocation to the United States. Her friends included Robert Oppenheimer, Victor Weisskopf, Max Delbruck, and Harold
Frey (Dash 1973:317). She was also a member of Enrico Fermi's research

team at Chicago (a corecipient of the Nobel Prize in 1963 with Eugene Wigner) whose innocent probing—*What about spin-orbit coupling?*—helped her solve the mystery of the nuclear shell structure problem. Rossiter pointed out that despite a lack of formal title, her inclusion in the Fermi group was essential for helping Goeppert-Mayer overcome her marginal status in science (1993:327–328). Further, with Hans D. Jensen of Germany, Maria codiscovered the shell model of the nucleus. However, of all her male supporters, her husband, Joseph Mayer, played the most important role in Goeppert-Mayer's professional life. Like her father, Joe Mayer expected, urged, and gave support for Goeppert-Mayer to continue her work in physics. He was probably among a handful of men in his time who did not expect their wives to occupy the traditional roles of women in society.

There are certain parallels between the Goeppert-Mayers and the Curies. In both couples, the husbands were *tremendously* supportive and understanding of their wives' aspirations and their commitment to science. Pierre Curie and Joe Mayer considered their wives more as allies and colleagues than simply companions and research associates. Most important, they respected their wives' talents and appreciated their brilliance. They were noncompetitive and willing to share credit equally with their wives for their joint work. Both Pierre Curie and Joe Mayer played an active role in their scientific spouses' professional lives. However, unlike Pierre Curie, Joe Mayer simply assumed the role of facilitator instead of collaborator. While women generally find marriage a distraction and children a burden, the unequivocal support from their scientific husbands made Marie Curie's and Maria Goeppert-Mayer's professional advancement easier.

Three women played a critical role in Rachel Carson's academic and professional life. Her mother was a mentor and champion throughout Carson's life. Despite financial hardships in the family, Maria Carson helped Rachel make her way to scientific writing. The person who had a decisive impact on Rachel Carson's professional writing was Grace Croff. A new assistant professor who taught freshman composition at the Pennsylvania College of Women (PCW), Croff encouraged and supported Carson's interest in writing. Mary Scott Skinker was another person who had an enduring impact on Carson's career. Had she not taken biology from Skinker at PCW, Carson would not have gone into science. Despite Carson's shyness, at Skinker's urging, Carson started a science club with a few science majors at the college and became its president.

Carson's coaches were women, while her sponsors and backers were men. Men opened the doors to scientific writing for Carson. At different stages of her career, each offered Carson useful and strategic guidance. If Mary Skinker was Carson's intellectual mentor, her supervisor at the Bureau of

Fisheries, Elmer Higgins, was her professional mentor. It was Higgins who reminded her of the limited possibilities for women making a career out of science. On the other hand, he was instrumental in helping Carson develop a career out of writing and marine biology. Specifically, Higgins steered her focus to writing on the sea and its creatures. During her 16 years of tenure at the Bureau working as a science writer, researcher, and editor, Carson developed extensive ties within and outside government agencies. Her sources represented a spectrum of individuals working in the government and universities, as well as business and industry (such as congressional aides, scientific experts, scholars in other fields, and journalists). She could rely on this network to find almost any piece of information she needed for her research on nature or the ocean. In addition to her literary agent, Marie Rodell, Carson's powerful allies included scientist Clarence Cottam, colleague Bob Hines, marine biologist William Beebe, editor Paul Brooks, and friend Edwin Way Teale.

Rita Levi-Montalcini and Barbara McClintock were both interested in biological science, but conducted research on different subjects—the former focused on the nerve growth factor and the latter on transposable genetic elements. Nonetheless, their networking strategies bear striking resemblance. This is probably due to a similarity in their personalities rather than work styles. Like McClintock, Levi-Montalcini formed lifelong friendships and later partnerships with a few influential male scientists. First, her college teacher became her mentor and later lifelong friend. She studied under Giuseppe Levi (no relation) at the Institute of Anatomy of the Turin Medical School. He was a role model for Levi-Montalcini in many respects. His passion for science, courage, and rigor set an example for his students. Three of his students eventually received the Nobel Prizes for medicine and physiology: Salvador Edward Luria (1969), Renato Dulbecco (1975), and Rita Levi-Montalcini (1986). Levi-Montalcini also sought advice from Luria and Dulbecco about her projects. Working in Levi's lab as an intern sparked her interest in a lifelong study of the central nervous system. This is how Levi-Montalcini described their evolving relationship:

> [Professor Giuseppe Levi] came to the conclusion that I was decidedly not cut out for research. I was automatically blacklisted as one of the students whom he defined as a "pain in the neck." . . . My lack of success convinced Levi that I was completely inept. . . . I was saved from this critical situation by having to be admitted to the clinic for an emergency operation. . . . When I returned to the institute, he allowed me to abandon the project and to start another topic. This one, unlike the two preceding ones, gratified me immensely and marked the beginning of a Master-disciple relationship, characterized by ever-increasing affection and reciprocal esteem which lasted until his death 31 years later. . . . This

research gave me my first glimpse not of the much-feared professor but of a master who had a real passion for his work, a critical sense far superior to that of the majority of biologists of the day (Levi-Montalcini 1988:59).

The training and motivation Levi-Montalcini received from Levi changed the course of her life (Levi-Montalcini 1988:60). Her admiration and affection for Giuseppe Levi parallels with Goeppert-Mayer's for Max Born. To a great extent, both Levi and Born became a father figure for their female protégé.

Neither McClintock nor Levi-Montalcini ever married. Marcus Rhoades (a graduate student at Cornell) became Barbara McClintock's closest friend (Comfort 2001:174–175), while Rodolfo Amprino, a fellow classmate, was Levi-Montalcini's confidant "during the most difficult years." According to Levi-Montalcini, Amprino was a gifted student and the best intern at Levi's lab. When she was having reservations about continuing her work after experiencing difficulties in the 1940s, Amprino pushed Levi-Montalcini to think long and hard about her career in research. When his inquiry of her research projects was met with silence, Amprino reasoned with Levi-Montalcini:

> One doesn't lose heart in the face of the first difficulties. Set up a small laboratory and take up your interrupted research. Remember Ramon y Cajal who in a poorly equipped institute. . . . did the fundamental work that established the basis of all we know about the nervous system of vertebrates (Levi-Montalcini 1988:89).

If Levi-Montalcini was fortunate to have Levi as her mentor in Italy, faith brought her to the most important ally in her scientific career. In the summer of 1940 in Italy, she read a scientific paper written by Viktor Hamburger, a renowned embryologist at Washington University in St. Louis, Missouri. She replicated his experiments in her bedroom with different results. Hamburger had also read results of her work published in Belgian and Swiss journals. He invited Levi-Montalcini to make a short visit to St. Louis to settle the dispute, which she accepted. The "invitation" and "expectation" were perfect examples of how the old-boy network works. Her correspondence with Salvador Edward Luria, a fellow classmate, paid off. She recalled, "Viktor had, in fact, asked him for references about me after reading my article, and it had been Luria's flattering recommendation that had prompted Viktor to invite me to St. Louis" (Levi-Montalcini 1988:138). A supposedly few months of visit in the United States was turned into 26 years of the "happiest and most productive years" of her life. More importantly, this three-decade long scientific partnership with Hamburger prompted Levi-Montalcini to make the United States her second home as well as a professional base. The support and goodwill Levi-Montalcini received from Hamburger and her American academic employer laid the foundation for her prize-winning research on NGF.

Stanley Cohen, a postdoc research fellow at Washington University who later shared the Nobel Prize with Levi-Montalcini, was another close friend and colleague. His expertise in biochemistry complemented her knowledge in neuroembryology, and she and Cohen worked personally together for years to identify NGF. Besides scientific skills, their differences in personality and work style contributed to a successful collaboration: "The complementarity of our competencies gave us good reason to rejoice instead of causing us inferiority complexes. 'Rita,' Stan said one day, 'You and I are good, but together we are wonderful'" (Levi-Montalcini 1988:163). Like many scientific couples, Cohen and Levi-Montalcini benefited from this professional partnership. In 1986, they shared the Nobel Prize in physiology and medicine for their discoveries of growth factors.

Similar to Mead, Carson, and Goeppert-Mayer, Levi-Montalcini also worked with leading scientists from other disciplines. Salvador Edward Luria and others introduced her to up-and-coming and renowned scientists such as James Watson, Tracy Sonneborn, and Hermann Muller. What is the significance of these intellectual exchanges for a female scientist? Through meetings with talented or eminent male scientists, Levi-Montalcini received affirmation of her work as well as her role in science. Together, these male scientists formed an image of what a scientist should be (Levi-Montalcini 1988:139). The interest of leading scientists in her work may have kept Levi-Montalcini going when she had reservations about her research. Throughout her career, she continued to cultivate ties within and outside the United States. Later in her career, she even managed to commute between St. Louis and Rome every six months by rotating with her young colleague and friend, Pietro Angeletti, as head of the two research labs set up in both countries.

Because her parents were often abroad and separated from their children, making contacts and seeking advice from others had been a part of Dorothy Hodgkin's upbringing. She succeeded in building extensive global contacts around the world that were related to her interests in science, education, and world peace. Very early on, Hodgkin understood the importance of ties on her scientific career in crystallography. When she could not get a job after graduating from Oxford, her childhood friend from Egypt, A.F. Joseph, came to her rescue. Joseph learned from a professor on a train about the possibility of a job with John Desmond Bernal at Cambridge. A supporter of female scientists, Bernal was conducting research on the use of X-rays to study biological crystals. This encounter prompted Joseph to contact T. Martin Lowry at Cambridge about the possibility of Hodgkin working with Bernal. Similar to Levi-Montalcini, word of mouth was how Hodgkin ended up working in Bernal's lab in October 1932 after college graduation. Similar to Hamburger's and Levi-Montalcini's situation, Bernal's offer to Hodgkin was an "invitation" and an "expectation."

As her mentor, Bernal did several things to enhance Hodgkin's scientific competence and reputation. He gave her a free rein in his lab. As a result, she often stumbled into interesting but unresolved problems. Bernal always put Hodgkin's name on his papers for which she had made some contribution. Gradually, she was able to build up her publishing record and garner attention from the scientific community. Through Bernal, she met many interesting people from other fields. Of this, she recounted:

> At 5 o'clock [Bernal] was due for a conference with Dr. Rosenheim at the National Medical Lab, so he took me with him and introduced me to everyone there. It was all very exciting—comparing results all round and planning what we should do next (Ferry 1998:86–87).

Dorothy maintained regular contacts with Bernal even after she left Cambridge. She often kept Bernal and his colleagues informed about her work and deferred to him on issues related to research and career. Besides having Bernal as her mentor and sponsor, Hodgkin developed and expanded her own networks. As Director of Studies of Somerville College for science students at Oxford, she was responsible for scheduling classes. This presented her an opportunity to develop ties with scientists across disciplines. Like other subjects in this study, she continued to enlist the help of others throughout her career. Most importantly, she knew where to turn to for advice or help. For instance, to acquire lab space and equipment for research, she approached Robert Robinson. A professor of organic chemistry supervising a large research group, Robinson offered to help. He even asked Hodgkin to give him a list of items she needed. Further, support from her husband, Thomas L. Hodgkin, in childrearing and other family matters allowed Hodgkin to concentrate on her work.

Rosalyn Yalow also found an understanding and supportive scientific spouse. Aaron Yalow, a graduate student in nuclear physics whom she met at the University of Illinois, approved of her academic and career goals. Yalow acknowledged the importance of her husband in accepting her not being a full-time homemaker (McGrayne 1998:339–340). Like Joseph Mayer, Aaron Yalow, a professor of physics, was a facilitator of his wife's work. Besides offering help and advice whenever she asked, Aaron Yalow critiqued every paper and speech Rosalyn Yalow wrote.

Rosalyn Yalow knew how to make the old-boy network work for her. Her experience in network-building was unique. She found women a hindrance to achieving her goals and men a resource for her career development. As she confessed, "In my career, I got help from men, not women. Neither of the two women professors I had in the physics department at Hunter did anything to get me along in physics" (McGrayne 1998:337). (Edith Quimby helped Yalow with good career advice, even if the Hunter College female professors did not (Straus 1998:68–70, 105–106).) Despite her aggressive style, Yalow had no problem working with male faculty. With the help of influential professors at Hunter—Herbert Otis, Duane Roller, and Jerrold Zacharias—she was awarded a graduate teaching assistantship at the University of Illinois. Yalow eventually had the opportunity to meet and study with eminent scientists like Maurice Goldhaber, who was Yalow's thesis advisor at Illinois. Goldhaber for Yalow was like Max Born for Maria Goeppert-Mayer and Giuseppe Levi for Rita Levi-Montalcini: They were willing to take bright female students under their wings.

Besides her husband, Solomon Berson was the second most important person in Yalow's career. Like Bernal, Berson was known to be very supportive of female scientists. Berson, as Viktor Hamburger did, teamed up with another female scientist to do research that eventually led to her receipt of the Nobel Prize. His reception to women's participation in science is best described by his closest scientific partner, Yalow: "It was a big surprise to me that a male physician would treat me as an equal. If Sol had not been that way, there would have been nothing I could have done about it" (Straus 1998:38). Berson's background in biomedicine complemented Yalow's training in physics, mathematics, and engineering. In addition to skills, their work styles complemented each other's. The Yalow/Berson scientific partnership bears striking resemblance to those of the Curies, Joliot-Curies, Mead/Bateson, and Levi-Montalcini/Cohen. All of these collaborations were characterized by a gender-division of labor, and yet each partner shared credit equally. Generally, the male member of the team was typically the "front man," dealing with the public and professional communities, while the female member focused and concentrated on running the lab or research tasks. To use the analogy of a business operation, instead of being codirectors, the male partner occupied the position of executive director and the female partner was the general manager of the research enterprise. Role-differentiation between male and female scientific partners allowed their research team to operate effectively and efficiently (Reskin 1978:14–15).

Yalow landed a job at the Bronx VA Hospital through the old-boy network. To survive and thrive in a male-dominated organization, she worked hard, became an integral part of the hospital, and won backing from influential male allies. Her strategy was effective (Straus 1998:83).

Though Yalow denied being discriminated against because of her gender, she was fully aware of the difficulties of fitting in at her workplace: "The VA was a male organization and it is very likely that without Sol I would always have been a second class citizen, because I was a nonphysician and a nonmale" (Straus 1998:83).

Ironically, it was probably due to her gender and nonmedical background that Yalow was perceived as less threatening to the VA organizational hierarchy. Despite Yalow's combativeness and competitiveness, she stood out among physicians and found a niche at the VA. Her presence and scientific talents were less likely to pose a genuine challenge to the gatekeepers. She might not have been content with her status and prospects at the VA, a major medical research center. However, her career trajectory at the VA would have been different if Yalow were a male physician. Two pieces of indirect evidence support this assumption. First, after working 18 years at the VA, her collaborator, Berson, moved on to Mt. Sinai School of Medicine to direct internal medicine. But Yalow had no desire to leave for better prospects. Second, after Sol Berson died suddenly in 1972, she renamed the lab "The Solomon A. Berson Research Laboratory." As its head, Yalow teamed up with Eugene Straus, a young male physician. For her, Straus became Berson's replacement in her research enterprise. In short, Yalow not only chose to stay at the VA to conduct research, she also continued to form new scientific partnerships with young scientists.

In many respects, Fay Ajzenberg-Selove was more assertive than Rosalyn Yalow in her dealings with employers. Aside from her wartime experience, this was due to the women's movement and affirmative action. These changes made it possible for women and members of other disadvantaged groups to challenge their employers in court. During her entire career, she used both the old-boy and new-girl clubs for employment and job changes.

In addition to her father, Fay Ajzenberg-Selove attributed her success to the support of two male scientists. She wrote Tom Lauritsen to ask him if he could hire her for the summer to work on an article rewrite. He got her a job at Caltech and became her mentor. Ajzenberg-Selove learned a great deal about nuclear physics from Lauritsen. They also collaborated for 21 years. She also got to know many influential male scientists, and became a part of their inner circle. Ajzenberg-Selove volunteered to type the notes of Eugene Wigner, a 1963 Nobel Prize recipient. At the house of MIT's Victor Weisskopf, she met many renowned physicists. Her old-boy network included notable scientists such as Herman Feshbach, Sam Goudsmit, and Denys Wilkinson. Unlike other female scientific pioneers, Ajzenberg-Selove was active in setting up a new-girl network. She traveled all over the world to participate in professional associations and meetings. For her, these meetings served both practical and political purposes:

If you work on interesting science, it is essential to shmooze with other scientists in your research field, and find out what they are doing at the time they are doing it. Your own work will not be current if you wait until your colleagues' results are published. . . . Attending meetings is essential. . . . The intense relationships between the people at scientific meetings . . . are built on personal connections, on past scientific alliances, and on rivalries. The latter are the results of struggles for recognition and for influence. The relationships between scientists are extremely intricate, and they are very important in determining whether they can do their research successfully (Ajzenberg-Selove 1994:137–138).

Ajzenberg-Selove's efforts in building gender-neutral networks were rewarded when her colleagues and friends, male and female, rallied behind her for research funding, salary increases, and in the tenure-appeal case.

It is true that the resources needed may not be always forthcoming, or the feedback received may not be always helpful. But one cannot deny that individual and collective efforts have both symbolic and practical value to each and every female pioneer: Pierre Curie's intervention on Marie Curie's behalf for the Nobel Prize nomination; Frederic Joliot's support to Irene Joliot-Curie; Franz Boas's mentoring of Margaret Mead; Marcus Rhoades's unwavering support for Barbara McClintock; the rally of Enrico Fermi and other prominent physicists behind Maria Goeppert-Mayer; the support from editor Paul Brooks and supervisor Elmer Higgins at the U.S. Bureau of Fisheries to Rachel Carson; Guiseppe Levi's mentoring of Rita Levi-Montalcini; John Desmond Bernal's sponsorship of Dorothy Hodgkin; the support from male professors at Hunter to Rosalyn Yalow; and Tom Lauritsen's mentoring of Fay Ajzenberg-Selove. In light of the rewards and recognition they received, this suggests that women's persistence, patience, and nurturing of important ties bears fruit.

As noted in previous chapters, it is generally difficult for women to find collaborators or mentors, especially male supporters. This assertion turned out to be untrue for subjects in this study. Without any contractual obligation or anticipation of reciprocation, why would these male scientists voluntarily do such a thing for female scientists? The evolving "agent-client" relationship between these male and female scientists cannot be explained by either the discrimination theory or the notion of homosocial reproduction. The behavior of these influential male scientists goes against the assumption of discrimination for women in predominantly male professions or the claim that individuals tend to hire or promote someone whose background is similar to theirs.

To explain this interesting paradox, I build on Rossiter's sympathy theory (1995:331–332), and argue that the evolving relationship is a form of human altruism in the scientific community. Social scientists have studied different forms of altruistic behavior in different historical, political, and cultural contexts (e.g., Monroe 1996; Oliner and Oliner 1988; Ortner 1999; Ostrower 1995; Wuthnow 1991). These compassionate or heroic acts ranged from people from all walks of life doing volunteer work for their community, to wealthy people donating large sums of money for social causes, to ordinary people risking their lives on behalf of others.

A considerable amount of human altruistic behavior is a product of culture (Allison 1992). If human altruism is a common occurrence in our daily living, why can't altruistic behavior develop in the institution of science? As gatekeepers of science, leading male scientists are in the position to offer professional support to the "underdogs," the "have nots," or the lesser-known scientists, such as female scientists (Epstein 1970:970–973). For example, Sir Frederick Hopkins, a founder of biochemistry and a Nobel laureate, was one of the few gatekeepers instrumental in promoting women's progress in science. Women occupied half of the research positions in his department at Cambridge. His hiring practice was unconventional at the time; none of the university departments had women on their research staffs (Chakravarthy et al. 1988:46–47).

Advocates like Hopkins were aware of the predicaments talented women face in a male domain. As noted in Chapter 3, based on their observations of or contacts with female students, spouses, or colleagues, they were often impressed with these women's ability to deal with challenges and setbacks. These female scientists set themselves apart from others not simply because of their scientific talents, but because of their indifference to obstacles and failures. Additionally, the dearth of women in science suggests that those who remained represented a highly select group compared to their male and female colleagues. Female scientists, as a group, made themselves more valuable in the scientific community; the impact of their work might have been disproportionate to their relative number. Perhaps for those reasons, they caught the attention of key male scientists in terms of performance and other characteristics. One can then argue that, in many respects, these female scientists were worth the gatekeepers' risk of offering help. As Pierre Curie observed of Marie Curie, "Women of genius are rare" (Curie 1937:125). When Robert Provine, Levi-Montalcini's postdoctoral fellow, was asked how he liked working with "Rita," his response was revealing: "Overall, it was a good ride. But sometimes it felt like working for Maria Callas and Marie Curie. . . . [However,] great ideas are a dime a dozen. The great scientist is one who delivers. And Rita delivered" (McGrayne 1998:219).

These female scientists were also keenly aware of the importance of maintaining contacts with everyone who assisted them in any way with their training and work. By doing so, they kept themselves fresh in the minds of those who might have been in the best position to help as their career progressed. This was why they did not hesitate to communicate new ideas and discoveries to influential male scientists. Regular correspondence was a golden opportunity to bring their ideas and work to the gatekeepers' attention. Another goal was to convince these powerful figures that they were a candidate worthy of serious consideration for sponsorship or mentoring.

This does not necessarily mean, however, that these female scientists lacked the knowledge of how to get along without any help. Yet the idiosyncratic behavior of Barbara McClintock and the effect of Solomon Berson's death on Rosalyn Yalow led many to doubt if they could remain professionally competitive. McClintock and Yalow were not content to occupy a marginal status in the scientific community. An analysis of their work histories reveals they continued to actively engage in advancing their careers. I argue that subjects in this study had resolved what Rossiter called a double bind between two contradictory social roles for women: (a) the permanently subordinate, helpful, almost invisible associates in the work of others and (b) the successful scientists who seek full professional recognition for their own discoveries and accomplishments (1982:268). No direct or indirect evidence suggests that any of these women perceived themselves less-than-equal to their male counterparts in terms of ability and status. Quite the contrary: their actions revealed that each considered herself as good, if not better than their male peers.

Those who do not subscribe to such an optimistic view might come to the same conclusion, but for different reasons. For political and personal motives, certain members of the scientific elite may be more eager than others to diversify the top of the scientific establishment. On the one hand, they aggressively lobby on behalf of a few token outsiders (such as women) for "window-dressing." On the other hand, these elites attempt to increase their "genetic diversity" to generate innovative research strategies. Recruiting people from different backgrounds into traditionally male-oriented science would help generate new ideas, the wellspring for the scientific community's continued survival and vitality. Historically, male scientists sought women for help in their work (Kohlstedt 1999). Edward Pickering, for example, hired many women to work for his Harvard Observatory projects because he found women's performance as computer workers satisfactory (Creese 1998:230–231). The research groups of William Henry Bragg and his son, William Lawrence Bragg, at Cambridge also produced a large number of successful female crystallographers (Rayner-Canham and Rayner-Canham 1998:68–75). Apparently, these practices are inconsistent with the notion of homosocial reproduction, when people tend to hire or promote someone who thinks and acts like them (Kanter 1977).

It is important to note that some female scientists and administrators also practiced homosocial reproduction. Dorothy Hodgkin, for example, hired one of her assistants on the spot. Eleanor Coller, the new computer specialist recruited by Hodgkin, recalled:

> The fact that [Dorothy Hodgkin] appointed me was typical in a lot of ways of how other parts of life were run. I just walked off the street, there had been no advertisement placed. I was the apple that had just fallen off the tree, and it was simpler to take that apple than to try and climb the tree to get a better one (Ferry 1998:310–311).

Some administrators and teachers at women's colleges prefer to hire and groom young female scientists with the prospects of becoming their successors (Lear 1998:45; McCay 1993:7; Rossiter 1982:18–19). With the exception of Irene Joliot-Curie, the experiences of female pioneers in this study run counter to the practice of homosocial reproduction in scientific communities. Many had influential male scientists as backers; they constitute examples of what I would call "heterosocial reproduction." Contrary to common belief, female scientists were able to enlist the help of influential male figures.

Additionally, there is support for Blalock's size discrimination thesis (1967), which holds that male resistance to women's participation increases when the number of women in the science workforce increases. Because women have not constituted a critical mass in the science population, they do not pose a real or perceived threat to men. Besides, a talented woman is generally viewed as less threatening than an equally talented man in cooperative or competitive settings. Women's participation on research teams headed by men helps alleviate the potential of masculine rivalry (de Meuron-Landolt 1975:152). This counterargument to the claim of "gender queue" is similar to the "harem effect" proposed by Rossiter, who observed that male scientists tend to favor women over men in hiring research assistants (1982:62). For a male scientist, having a group of young female assistants in a lab is perceived as less threatening compared to having an equal number of male assistants in the same setting. Another reason why male scientists prefer to hire female associates is that women tend to be less geographically mobile when compared to men; a low turnover rate of female research associates minimizes the costs of finding replacements as well as the investments in training.

There are important implications of actions by female scientific pioneers and their alliances. For one, the individual actions of these female pioneers challenge the assumption about female scientists that they are isolated and cannot easily develop professional ties. These women demonstrated that they could engage in a complex web of collegial relationships in the scientific community. Another significance of these actions was that, ironically, they reinforced the traditional gender stereotypes in society where men typically are the defender of women's interests, and women are in need of men's help.

## (NON)LINEAR CAREER PATHS TO SCIENTIFIC SUCCESS

The experiences of female scientific pioneers provide contradictory evidence for the linear career-path model[7] and indicate there are alternative paths for women to make their scientific careers distinctive. Subjects in this study used different methods to get ahead and stay ahead. Additionally, there was no evidence showing that one career model was more effective than others. The assertion that marriage is incompatible with work was not supported in the case of these female scientific pioneers. They chose or followed a particular path in light of their own personalities, outlooks on life, and circumstances.

One of the prevailing assumptions is that in order to succeed in science, a practitioner is expected to follow *the* linear path without any interruptions and to devote to work to the exclusion of personal life. I label this ideal of scientists working zealously in their offices or labs from daybreak until midnight as the "7 to 11" approach to scientific work. This work style is incongruent with gender roles. Traditionally, society expects men to devote themselves to careers and women to family commitments. Proponents of this approach could use the career experiences and career outcomes of renowned male scientists to underscore the significance of following this path to career success (e.g., Einstein, Feynman). Even skeptics of the male model of science have to admit that very few female scientists have been able to deviate from the linear career path to achieve career success (Schiebinger 1987:314). For women, those who have become successful *and* able to balance work and family are rare. If this is the case, the experience of female scientific pioneers in this study was an anomaly. Many had relatively stable and satisfying marriages and, therefore, could be ideal role models for professional women. Of the ten female pioneers in this study, seven were married, and with the exception of Fay Ajzenberg-Selove, all had children (Marie Curie, Irene Joliot-Curie, Margaret Mead, Maria Goeppert-Mayer, Dorothy Hodgkin, and Rosalyn Yalow). Consistent with the expectations of successful female scientists, these seven women all had scientific spouses who were in the position to provide emotional and/or professional support to their wives' careers. All, except Mead, stayed married to their spouses throughout their careers, a phenomenon rather uncommon among professional women. The other three female pioneers, Barbara McClintock, Rachel Carson, and Rita Levi-Montalcini, remained single by choice. An optimistic reading of these statistics suggests that (a) conformity to the male model of science was not a prevailing practice among the subjects in this study and (b) there are role models of *successful* married female scientists. It could be argued that these results clearly challenge the image portrayed in the media of female scientists as "deviants" (LaFollette 1988).

Most intriguing is that virtually all of them were aware of the demands of work and home, yet succeeded in making their mark in science whether they were single or married. The life- and career-experiences of these women of the prewar and postwar generations suggest a pattern of multiplicity as opposed to uniformity. The fact that married women as well as single women with family responsibilities (e.g., Carson) managed to have successful careers in science does not nullify the debate over incompatibility between career and family for women. Quite the contrary: the experiences of these successful female scientists underscore the demands of career and home.

To fulfill their professional and family obligations, female scientific pioneers followed nonlinear career paths to science. Instead of putting work above family or vice versa, they divided their time between science and family almost exclusively, with little or no time for themselves. Having a scientific spouse may not always help a woman's career. However, almost without exception, these women minimized conflicts between professional and gender roles by having primary and secondary caregivers for childcare and housekeeping—a supportive spouse (or parent and relatives) and external help (Pycior et al. 1996; White 1982). Apparently, they did not have the traditional marriages, where the wife is expected to put her family and husband's career above and ahead of everything else, and that she usually occupies a subordinate role in the household. These female pioneers all had egalitarian marriages, in which the spouses had similar commitment to their work and shared equal power in the household (Pycior et al. 1996:248).

For both the husband and wife, marriage does not and should not end the wife's commitment to scientific work. Although marriage and career may not be entirely compatible, they are not mutually exclusive domains. Scientific marriages may at times be crucial to a woman's career success in science. In this study, examples of the tremendous impact of a supportive scientific husband on the wife's career ascension include the marriages of: (a) Marie and Pierre Curie, (b) Irene Joliot-Curie and Frederic Joliot, (c) Margaret Mead and Gregory Bateson, (d) Maria Goeppert-Mayer and Joseph Mayer, (e) Dorothy and Thomas Hodgkin, (f) Rosalyn and Aaron Yalow, and (g) Fay Ajzenberg-Selove and Walter Selove. However, the career trajectories of both married and single female scientific pioneers in this study do not support the claim that marriage seems to be positively related to greater professional activity.

In addition to luck, these women were so disciplined and efficient that they could perform the multiple duties of being a professional and a spouse/mother (or daughter). It is also imperative to point out that these female pioneers successfully juggled between two different roles *almost* to the exclusion of personal life. Thus, critics of the "linear career path to science" can hardly make the claim that the alternative nonlinear career path to science is a model for women to emulate. Meeting the demands of career and home was an issue for subjects in this study. Family demand was factored into their choice of remaining single or having a family. The outcome of their cost-and-benefit analysis was reflected in their decision of marrying or not. Some female pioneers managed to modify their lifestyle to accommodate family into their career and lived up to the traditional expectations of being a woman and a professional. This is a challenge to the assumption that women should take a conciliatory approach to achieve success in science. For some women, doing pioneering scientific work and having a family are not mutually exclusive goals, but are also not without a price.


## SUMMARY


In this chapter, I argue for the existence of norms and counternorms in the scientific establishment. The discussions reveal the shortcomings of the normative structure of science and the kinds of career development and advancement it fails to explain. Not all norms in science are gender-blind. Women in particular have difficulty in adhering to these norms. The life and work of the women in this study demonstrate how extraordinary women reflect, or depart from, cultural norms. Collectively, these (counter)norms positively and negatively affect the career development and advancement of female scientists. As a result of trial and error, female scientific pioneers successfully navigated through the male culture of science.

An interesting question emerges from this chapter: If counternorms exist, what happens to male scientists who follow them? If each of the ten women in this study were male, would each have met with the same level of success? We do not know, but one can argue that the obstacles these women faced might have pushed them to work harder, and thus led to the development of strategic and strong professional relationships with powerful male scientists in the field.

Science is a social activity and a social product. There is both *competition* and *collaboration* in the scientific world between male and female scientists. Practices of *homosocial reproduction* and *genetic diversity* are evident in the production of female scientific pioneers. Within the matrix of science, women's prospects can be affected by *accumulative disadvantages* as well as *accumulative advantages*. Unlike most female scientists who are often found in the periphery of the scientific establishment (*the outer circle*), these female scientific pioneers were located in the core of the scientific establishment (*the inner circle* or *the invisible college*). Contrary to the popular perception, there was no specific career path to scientific success for them. Female scientific pioneers followed the linear *and* nonlinear career paths to get ahead in science. The majority upheld the view that marriage is compatible with career (Curie, Joliot-Curie, Mead, Goeppert-Mayer, Hodgkin, Yalow, and Ajzenberg-Selove), yet some did not (McClintock, Carson, and Levi-Montalcini).

Many female scientific pioneers received much encouragement and support from key male figures in their lives, beginning with their (grand)father. They had an understanding that self-promotion was essential for advancing in science. For the most part, their networking approach was voluntary, inclusive, and crossdisciplinary. From learning to collaboration, they chose scientific spouses and collaborators who shared the same purposes, values, and goals. Additionally, they associated with people with high status and strong resources in the scientific community. They invested in forming lifetime networks with people who strove for the same vision and goals in life and career.

The most important of all is that subjects in this study, with or without a scientific spouse, found ways to sustain their scientific productivity. It was common for these women to have domestic help from in-laws, parents, spouses, or outside hires. Thus, succeeding in science is truly a social activity in light of the norms and counternorms in the scientific community. Unfortunately, the significant efforts of networking and charting alternative career paths among these extraordinary women hardly altered policy decisions about women's participation in science at the national or global level.

*Chapter Seven*

# What Lies Beneath?

What the subjects in this study did and achieved is a paradox. They had fewer resources and opportunities to make it in science, yet were able to become pioneers in the scientific world. This concluding chapter gives a summary of analyses presented in the book. Aside from synthesizing the findings and assessing the implications for policy debates, theoretical development, and research, I speculate on the prospects of women in science: What does the future hold for women in science? What can be done to improve their current status in science? Finally, I consider implications of the findings for women in other domains. The discussions lay the groundwork for comparative studies in other endeavors.

## SUMMARY AND DISCUSSIONS

Achieving career success in science presents a challenge, especially to women. The culture and norms in society and science have affected men and women differently, primarily because of gender socialization and the structure of the scientific profession. The career experiences and career outcomes of the ten notable female scientists featured in this study set them apart from other female scientists as well as from those who departed from science. One can hardly deny that these women became pioneers in their field because they circumvented numerous obstacles to make significant contributions to the field and the profession. By studying female scientific pioneers, we can see these micro and macrolevel forces more clearly.

The ten women in this study may well have characteristics that are not representative of the general scientific population. How did they become great women of science in the eyes of their colleagues, and others did not? In addition to being extremely talented, highly motivated, and hardworking, they benefited from major cultural, economic, and political developments. These changes broke down traditional barriers for women to education and employment. In addition, these women managed to build and sustain a vital support mechanism for personal growth and professional development. Support from family, schools/research institutes, and the scientific community was indispensable throughout their careers. Simply put, they had the confidence, talents, and necessary contacts to attain great accomplishments. Under the sponsorship of or in partnership with a few key figures in the scientific establishment, typically a male related by marriage or by mentoring, they succeeded in joining the ranks of scientific pioneers.

The experiences of female scientific pioneers also tell us that the ideology of meritocracy in science offers a very simplistic view of how the scientific community has operated. The fact that these female scientists faced numerous formal barriers in their pursuit of pioneering work highlights the significance of nonmeritorious norms and unfair practices in science. As these women's experiences illustrate, to move up the hierarchy of scientific establishment, women must know how to navigate within this male domain.

Among the factors facilitating women's career development, the ability to establish and maintain ties with gatekeepers in science is most critical. Contrary to the assumption that female scientists tend to work in isolation, these female scientific pioneers worked hard to build up their networks. However, it would be a mistake to attribute their career success in science primarily to their ability to netvigate. Many actually took bold and strategic actions to become part of the old-boys club.

Female scientific pioneers took different paths to achieve success in science. Some followed the traditional linear career model and devoted their time almost exclusively to science, while others managed to engage in science and have a family. Nonetheless, regardless of their lifestyles, these women's experiences and the strategies they adopted to handle the demands of work and home highlight the conflicts between science and family for women.

Female scientific pioneers present a unique opportunity to revisit stratification and mobility in science. We may be able to learn much more from understanding why some female scientists who begin what look like very different career prospects end up being pioneers in their field.[8] Such a study will make a significant contribution to public policy issues as well as to our understanding of the problems and opportunities for women in science. For one, their career experiences and insights might provide clues to policymakers in devising ways to enhance the careers of young women in predominantly male scientific fields. In addition, identifying factors producing female scientific pioneers could help women to become successful in the fields of science. Because of women's underrepresentation in science, society needs far more than just the prospective female Nobel laureates to stay in science to meet the increasing need for a scientific workforce.

This book presents ten female scientific pioneers as the way they were, not the way all women who aspire for scientific success ought to be. There is no specific career path to becoming a pioneer in science, and the paths to success differ among subjects in this study as well. Their career development was by no means uniform or predictable. Some were highly competitive, others highly cooperative, and others a mixture of varying degrees. Their career advancement process also reveals amazing simplicity and vast complexity. Adopting the earliest forms of scientific enterprise, many female scientific pioneers formed successful scientific partnerships with brilliant people (predominantly men) through marriage or professional collaboration (see Table 3). Most importantly, three quarters of the partnerships were egalitarian in terms of actual (rather than perceived) credit-sharing between female pioneers and their collaborators. The rest were either paternal or maternal types of partnership, primarily due to age differences between subjects in the study and their collaborators. As shown in Table 4, most of these collaborations were cross-gender and cross-cultural. However, the underlying processes resulting in the positive impact of different forms of collaboration may vary for male and female scientific pioneers.

## POLICYMAKING

What is the impact of so few female scientific pioneers on women in science and on society in general? According to Straus:

> [These female scientific pioneers] have forced [themselves] to the very top of the man's world. It had not been done before, not in the world of medicine [and science.] From [their stories] we can learn about our [society's] values, about our scientific establishment, and about our own ambitious impulses. If we fail to use the lessons of trailblazers like Yalow [and others] then we all have a harder way to go. . . . [Their] careers can guide us to understanding more profoundly the plight of women in science and society (1998:xiv, 6).

Others, such as Ochse, also underscored the need for bestowing honor and recognition on exceptional scientists:

> Excellence breeds excellence. . . . a few outstanding creative people in a culture . . . may act as models, setting high standards and showing others in the field

**Table 3.  Types and Forms of Partnerships**

| Partners | Legal Marriage | Professional | Paternal: Master–disciple | Maternal: Master–disciple | Egalitarian: Credit distribution |
|---|---|---|---|---|---|
| Marie Curie & Pierre Curie [Co-recipients of Nobel Prize] | **Yes** | **Yes** | No | No | **Yes** |
| Marie Curie & Irene Joliot-Curie | No | **Yes** | No | (Mother/Daughter) **Yes** | No |
| Irene Joliot-Curie & Frederic Joliot [Co-recipients of Nobel Prize] | **Yes** | **Yes** | No | No | **Yes** |
| Margaret Mead & Reo Fortune | **Yes** | **Yes** | No | No | **Yes** |
| Margaret Mead & Gregory Bateson | **Yes** | **Yes** | No | No | **Yes** |
| Maria Goeppert-Mayer & Joseph Mayer | **Yes** | **Yes** | No | No | **Yes** |
| Rita Levi-Montalcini & Viktor Hamburger | No | **Yes** | No | No | **Yes** |
| Rita Levi-Montalcini & Stanley Cohen [Co-recipients of Nobel Prize] | No | **Yes** | No | No | **Yes** |
| Dorothy Hodgkin & John Desmond Bernal | No | **Yes** | **Yes** | No | No |
| Rosalyn Yalow & Solomon Berson | No | **Yes** | No | No | **Yes** |
| Rosalyn Yalow & Eugene Straus | No | **Yes** | No | **Yes** | **Yes** |
| Fay Ajzenberg-Selove & Tom Lauritsen | No | **Yes** | **Yes** | No | No |

**Table 4. Types of Collaboration**

| Cross-gender Collaboration | Cross-cultural Collaboration | Same-gender Collaboration |
| --- | --- | --- |
| Marie Curie & Pierre Curie | Marie Curie & Pierre Curie | Marie Curie & Irene Joliot-Curie |
| Irene Joliot-Curie & Frederic Joliot | | |
| Margaret Mead & Reo Fortune | Margaret Mead & Reo Fortune | |
| Margaret Mead & Gregory Bateson | Margaret Mead & Gregory Bateson | |
| Maria Goeppert-Mayer & Joseph Mayer | Maria Goeppert-Mayer & Joseph Mayer | |
| Rita Levi-Montalcini & Viktor Hamburger | Rita Levi-Montalcini & Viktor Hamburger | |
| Rita Levi-Montalcini & Stanley Cohen | Rita Levi-Montalcini & Stanley Cohen | |
| Dorothy Hodgkin & John Desmond Bernal | | |
| Rosalyn Yalow & Solomon Berson | | |
| Rosalyn Yalow & Eugene Straus | | |
| Fay Ajzenberg-Selove & Tom Lauritsen | Fay Ajzenberg-Selove & Tom Lauritsen | |

how to meet those standards. . . . Societies breed excellence in their members by publicly acclaiming and rewarding those who create something excellent" (1991:335–336).

In short, the scientific community would not be able to attract and retain the best and brightest if individuals with high performance were not properly rewarded.

Does the success of female scientific pioneers reflect the success of socialization of scientists? That is, as a group, do they live up to the ideal of a scientist? Some think not. For example, Barber (1995) underscored a need for broadening the pool of science and engineering workforce through diversification:

> We acknowledge the problem, we expend the resources, dedicated people exert enormous effort, yet somehow we cannot produce the desired result. This conundrum must be addressed, if we are to avoid further waste of valuable resources. . . . Transforming the culture of science is the key to narrowing the science and engineering gender gap. Rather than assimilating women to existing standards, interventions must focus on broadening the cultural norms of the profession. A more diverse culture of science would comfortably support a more diverse group of scientists, allowing real progress toward equity. . . . [I]ncreasing diversity could have the added benefit of stimulating creativity in science and engineering overall (Barber 1995:229, 232).

The phenomenon of heterosocial reproduction observed among the subjects in this study can be considered as a positive step in the right direction.

Gender has been considered as a factor in the making of scientists. The scientific community is characterized by a male-oriented culture and a hierarchical structure. The case of female pioneers suggests that the scientific community is subject to two conflicting goals: supporting the norms of universalism and disinterestedness and maintaining the status quo. To survive, gatekeepers of the scientific community must adapt to changes and be open to talent. On the other hand, they would like to maintain male dominance at the top. Policymakers should explore particular problems women may have at the elite level of the scientific establishment. This case study of mobility in science is one example of why policymakers should begin to look more closely at the differences between men and women in their rise to the upper echelon of the scientific community. For many years, public attention on female scientific elites has been lacking, and women have been excluded from the analysis of scientific elites. It is hopeful that this trend has begun to reverse as studies on women with successful careers gain momentum. A new way of looking at the similarities and differences between men and women would have implications for the training of scientists.

The current "tournament model" in science has far-reaching implications for the production of scientific pioneers. According to Rosenbaum, careers are conceived as a sequence of competitions—the winner advances to the next level for further competition, while the loser is denied the opportunity to compete for high levels. The outcome of each competition has implications for a person's chances for mobility in subsequent selections (Rosenbaum 1984:42). Yet this system might eliminate some highly talented young scientists, male and female, from advancing to the top. Rosenbaum's observations of the challenges that men and women face in the workplace may also apply to new generations of scientists:

> At a time when men are increasingly more involved with child-care and household responsibilities and women are more career-oriented, the timetable imposed by the tournament forces the career system to neglect important segments of the workforce. . . . The tournament system makes its most important selections —and consequently its greatest demands—at precisely the period when individuals customarily have had the greatest demands from their family responsibilities. . . . These selection systems are making fateful decisions about employees at precisely the time when employees are relatively less free to immerse themselves fully in their work (Rosenbaum 1984:286, 298–299).

Such a selection system could hamper individuals' careers and the growth of scientific knowledge; young scientists are required to demonstrate their abilities and achieve their best performance at a time when they are most pressed for family and domestic responsibilities. Again, Rosenbaum's characterization of tension between demands of family and career reveals the inherent weaknesses of the tournament system in picking the best and brightest:

> Rather than selecting the most able employees, early selection systems may be selecting those employees who have the least demands from, or least concern about, their families. By making its demands coincide with the period of greatest family responsibilities, the organization wins only if the family loses and the organization does not win employees who are strongly committed to their families (Rosenbaum 1984:298–299).

The study of female scientific pioneers reveals that it is possible for individuals who are highly committed to family to pursue pioneering work; they all had reliable childcare assistance from relatives or hired employees. In addition, adjustments or changes in the selection and training of scientists could improve the pool of potential candidates for joining the ranks of scientific pioneers.

A related question is how many pioneers a society really needs. Ideally, society should produce as many as it possibly can. The preceding discussions suggest that society would be better off if we considered alternatives to the tournament-mobility model. This case study of female scientific pioneers challenges policymakers to think long and hard about promoting scientific literacy for all, which is the current theme of much science education. Based on the educational experiences of subjects in this study, we should be concerned about developing scientific literacy in all students instead of putting resources into developing relatively few individuals who do pioneering work. The educational system must be changed so that more people from different socio-economic backgrounds can participate in science (Hanson 1996; Preston 2004; Rosser 2004). Specifically, educators need to experiment with new ways of teaching in and outside the classroom so that more people become creative thinkers or successful scientists (e.g., Eisenhart and Finkel 1998). Additionally, social scientists should pay more attention to elements of an individual's life, such as family upbringing and parental involvement in education, to improve the chances of successful outcome (e.g., Bengtson, Biblarz, and Roberts 2002; Massey, Charles, Lundy, and Fischer 2003). An analysis of the upbringing and academic experiences of subjects in this study tells us that family and educational environments can be manipulated so that the development of certain characteristics can be fostered among individuals.

## THEORETICAL DEVELOPMENT

There is good evidence to support a major rethinking of the norms and practices in science. Science is both a personal and communal activity; results of the analysis suggest the inseparability of science and society. Therefore, we should view and interpret scientific achievements in the social milieu that affect life and work. As Comfort puts it, "Science cannot be understood without placing it in a social context. . . . [T]he social context cannot be understood without grounding it in the science" (2001:12). Subjects in this study seemed to have similar experiences as they moved closer to the top of the scientific establishment, supporting the idea that there should be a uniform, dynamic approach to understanding the production of female scientific pioneers. Specifically, researchers should adopt a developmental view. This would allow them to detect any similarities or differences in various stages among female scientific pioneers.

It is imperative to bring individual, structural, and institutional forces into analysis of scientific careers. Results of this study reveal a multidimensionality of social influence. There is *no* support for the suggestion of chance or luck. The career success of subjects in this study was a consequence of interaction between individual and situational factors over a substantial time period. Besides talents and abilities, performance/achievement/success in science was context/setting dependent. These ten women did not enjoy the same opportunities their male counterparts did on either an individual or a collective level. That is why most people did not have illusions about their pursuit of pioneering work. But their scientific achievements suggested that does not mean anything. These female scientists became pioneers in their fields because they deserved to be.

There is no way of knowing if their levels of achievements and recognition would have been any different had their formal and informal barriers been lifted at the outset. Results of this study also bolster the homosocial theory of sex roles. Female scientific pioneers had male supporters, who were usually powerful or resourceful. Being part of the old-boy network gave these women indirect access to limited-but-valuable resources for educational and career advancement. Ironically, their participation in the invisible college reinforced and perpetuated the traditional gender hierarchy in science. Their acquiescence may have perpetuated existing practices in the scientific establishment. Most of them relied on individual efforts rather than collective actions to overcome perceptual and political barriers to women. We do not know if and how these adaptive strategies affected women's progress in science. Some scholars have pointed out that actions of this kind are more likely to set the clock back instead of forward. For instance, Bradford (1980:42) argued for changes in existing social-control mechanism:

> First-born daughters who have followed careers and deferred children cannot be thought of as advancing the feminine cause, as we have become 'honorary males.' We have too often accepted the male-dominated view of society, and we are frequently impatient of women's inability to change their lot. . . . This should require a major re-organization of the economic and social roles of men and women.

What is striking is that virtually none of the subjects in this study (except Fay Ajzenberg-Selove) felt as though they were discriminated against based on their gender. Most of them refused to acknowledge gender bias or discrimination in science. The aspirations and careers of these female scientific pioneers revealed sharp contrast to traditional gender socialization. Because of women's marginal position in science at the time, these women might have adopted the views of members of the majority. Denying the existence of gender discrimination might have lessened their sense of insecurity in a male domain (Stolte-Heiskanen 1983:81).

This observation raises the question of how would-be pioneers who did not make it like these female scientists did feel about their gender role in their career. There are several implications of this observation. For one, women in positions of influence or power might not be willing to assist women who are "on their way up." The reason is that some women in positions of power might want to remain the only woman in that position due to prestige, reduced competition, and novelty. Further, a female pioneer might be reluctant to help an up-and-coming female scientist, because the pioneer believes that the up-and-coming scientist needs to find a male scientist to help her, not a female pioneer. As a result, the traditional model of how a woman becomes a pioneer is perpetuated. Hopefully, the involvement of female scientists such as Yalow and Ajzenberg-Selove in mentoring might slowly alter this perception.

Additionally, the findings lend support to two different versions of the historical effects hypothesis proposed by Rosenbaum (1984:181)—the *Horatio Alger* version and the *cumulative privilege* version. The Horatio Alger version refers to the phenomenon that people who manage to overcome obstacles of lower-status origins *and* rise quickly within the first few years of employment are generally seen as more capable. As a result, they advance quickly when compared to their coworkers with the same level of attainment at the early stages of their career. Simply put, individuals who overcome initial disadvantages are deemed more able and are given more opportunities for advancement. The ascension of Marie Curie, Rachel Carson, Rita Levi-Montalcini, Rosalyn Yalow, and Fay Ajzenberg-Selove to the upper echelon of the scientific establishment fits the description of this version of *negative* historical effects. In contrast, the cumulative privilege version postulates that people who enjoy higher-status positions are given more opportunities for further advancement than others are at the early stages of their careers. Irene Joliot-Curie, Margaret Mead, Barbara McClintock, Maria Goeppert-Mayer, and Dorothy Hodgkin meet the description of this version of *positive* historical effects.

## RESEARCH ON STRATIFICATION AND MOBILITY

There is a growing amount of literature on women in science to shed new light on stratification and mobility in science. In addition to works from the sociology of science and the history of science, we can shed light on mobility in science using literature on work and occupations. Recent studies of scientific pioneers represent a narrow view of mobility in science in general and scientific pioneers in particular. Current works on eminent scientists offer an incomplete account of scientific pioneers. The descriptions and discussions have been generated almost exclusively from or about male scientists. As a result, we do not know whether and to what extent the conclusions drawn can be applied to female scientific pioneers. This retrospective study fills this void in the literature by including female scientific pioneers from different fields and contexts. However, we should consider both the benefits and dangers of counting female pioneers out. In addition, when we search out pioneers, we should avoid, where possible, using nineteenth- and twentieth-century criteria to define scientific pioneers.

The small number of female scientists among pioneers may be due to the way in which the "scientific pioneer" is identified. We could revisit stratification and mobility in science, for example, by revising the means by which the scientific accomplishments of women have been measured and evaluated. If women have been historically excluded from participating in science via education, employment, and holding positions in mainstream scientific societies, it would be difficult to give full credits to female scientists who truly deserve them. As pointed out by Long, Allison, and McGinnis (1979:817), misappropriation of credits is not only unfair to individual scientists, it also hampers scientific progress.

One should look beyond the existing selection criteria *and* look harder to seek out *women among the best*, not *the best among women*. There is a possibility that their exceptional scientific achievements are in part a function of differential attrition rates. The relatively high attrition rate of women from science leaves us with a more highly select group of women in science. Therefore, the definition of "female scientific pioneers" should be broadened to include not only those who have received a Nobel Prize or scientific honors, but also to include women doing good work, helping others, or making the world a better place. Not surprisingly, subjects in this study meet all of these criteria.

Quantitative studies of scientific careers have not paid enough attention to the impact of interplay between individual, structural, and institutional forces. The forces that facilitate the career success of female pioneers may be different from those for the general scientific population. Male and female pioneers in science may have to overcome similar and yet different sets of barriers. Those who seek to understand the production of scientific pioneers may benefit from a multimethod approach, which has been used in recent studies of work and occupations (e.g., Blair-Loy 2003; Hodson 2001; Mortimer 2003; Moss and Tilly 2001). These studies relied on both quantitative and ethnographic data to give a richer description and analysis of the possibilities and problems that workers, including teenage workers, face in the labor markets. Drawing on biographies and ethnographies, we can pinpoint general and specific factors as well as historical and personal events contributing to successful pioneering work by female scientists. Equally important, this qualitative data allows us to gauge the intensity of their experiences at different stages of their careers. Potentially important information can be gained if future analyses use existing biographical and ethnographic data about notable female scientists. A concrete and well-documented analysis of these women's educational and career progress could enhance our understanding of mobility in the scientific establishment.

A more ambitious goal is that this data may provide useful suggestions as to what can be done to improve the career achievements of women in science. Though in no way representative of the population of female scientists, qualitative analyses of a statistical rarity such as female scientific pioneers allow us to revisit stratification and mobility in science from a different dimension. Would female scientists today feel they are being neglected, not being given all the tools to conduct good research and, therefore, not making the list of award-winners and/or become big-grant recipients? Are the disadvantages that blocked these ten women still working today? These issues are explored in detail later in the chapter.

Not enough attention has been paid to the investigation and analysis of the network-building of female scientists, even though their personal and professional networks are intertwined. Succeeding in science, for women, is beyond talents and merits. Today's research in many fields tends to be large-scale, crossdisciplinary, and team-oriented. The need for expensive equipment, support staff, and research space, for example, requires female scientists to rely more on networks than ever. This could open up more opportunities to engage in cross- or same-gender collaborations within and outside their fields.

We should also consider the implications of findings in this case study of scientific pioneers for men and women in the corporate world (Kanter 1977; Powell and Graves 2003). This book examines individual mobility in a single male-dominated profession. Findings may have relevance for other types of professions or settings. This case study gives us a glimpse of the leadership skills of female scientific pioneers. Subjects in this study were strong leaders *before* they entered science. They were strategists and "wore many hats." Throughout their lives and work, they prepared themselves to be *apprentice* (to receive formal and informal training from family, peers, teachers, and colleagues), *diplomat* (to obtain help or support from those who could provide help or support), *cheerleader* (to deal with personal and professional setbacks), both *good cop* and *bad cop* (to handle conflicting work and family demands), and even *therapist* (to cope with political and social upheavals). Doing scientific work and being a leader are not mutually exclusive tasks. Seven of the ten subjects in this study had extensive administrative or leadership experience:

1. Marie Curie founded and became the head of the Radium Institute in Paris. She also directed a large-scale radiology service in France during the First World War (1914–1919).

2. Irene Joliot-Curie succeeded her mother and became director of the Radium Institute (1946–1956). She also served as the Undersecretary of State for Scientific Research (1936) and Commissioner for Atomic Energy (1946–1951) in France.
3. Margaret Mead was the president of the American Anthropological Association (1960) and the American Association for the Advancement of Science (1975).
4. Rachel Carson was promoted from junior aquatic biologist to editor-in-chief at the U.S. Bureau of Fisheries (1936–1952).
5. Rita Levi-Montalcini was a codirector of two research labs: the lab at Washington University, St. Louis, and the Center for Neurobiology in Rome (1961–1979).
6. Rosalyn Yalow directed the Solomon A. Berson Research Laboratory at the Bronx VA Medical Center (1973–1992).
7. Fay Ajzenberg-Selove chaired the Division of Nuclear Physics of the American Physical Society (1973–1974), and the Commission on Nuclear Physics of the International Union of Pure and Applied Physics (1978–1981).

These examples illustrate that the success of female scientific pioneers is not only limited to making discoveries. They are also shrewd managers or leaders in their fields or organizations.

## THE ALBERT EINSTEINS AND MARIE CURIES IN THE NEW MILLENNIUM

Becoming a pioneer is the pinnacle of one's scientific career achievements.[9] Are there similarities and differences in the making of male and female scientific pioneers in their early or later years? This important question has received minimal attention in studies of scientific pioneers. As Fox and Stephan observed, "Variations for women and men in scientific careers are significant because of the relationship between gender and science: science both reflects and, in certain ways, exemplifies gender stratification within society" (2001:110).

There are parallels between female scientific pioneers in the twentieth century and men in science observed by Zuckerman (1988:530). Highly successful male scientists studied under great masters or had powerful sponsors. Additionally, the stratification system of science is neither wholly universalistic nor wholly particularistic. Particularistic standards are applied early in one's career, even before a person has a chance to show his performance.

Specifically, sponsors would carry out informal assessments of promise or potential in allocating resources and rewards. As time goes by, there is a diminishing influence of sponsorship and prestige of the department granting the doctorate on career progress. Research performance as measured by publication and citation would increase its significance in the allocation of resources and rewards.

There is also support for the claim that recruits into the upper echelon of elites in the scientific establishment are highly selective in terms of family-class background (Zuckerman 1996). With the exception of Rosalyn Yalow, all female scientific pioneers in the study came from either professional or privileged backgrounds. This finding corresponds to observations made for the academic profession (Crane 1967), the power elite (Zweigenhaft and Domhoff 1998), and the trustees of arts boards (Ostrower 2002).

In a recent study of two arts boards, Ostrower underscored the dual influence of class and organization on diversifying board membership. Organizations are self-perpetuating entities. Their membership composition and recruitment practices, Ostrower noted, reflect shifting organizational and environmental changes. When gatekeepers recruit new members, they generally look to individuals of similar class-background. As Ostrower pointed out, even when board members intend to diversify the composition of the board, they select candidates who can bring in new sources of support and perpetuate the prestige and influence of the board (2002:15). Given the exclusiveness of the board, under what circumstances would members from different backgrounds be admitted? It is plausible that a person who has ties to the board and whose background (being a woman or a racial/ethnic minority) would give the board access to the communities or groups the board hopes to reach (women or minorities). Despite gender and/or racial differences, new entrants to arts boards share commonalities with incumbents in terms of class background or intellectual interests (Ostrower 2002:48). In other words, potential candidates or prospective members for elite status are expected to meet the old-boy standards. This led Ostrower to conclude, "Board membership is highly coveted and the social and cultural barriers to membership are substantial" (2002:110). Thus, this process of recruitment will not identify individuals, regardless of their background, who are not part of the board's network (Ostrower 2002:54). This observation underscores the significance of netvigating for women aspiring to scientific success. Nonetheless, it remains to be seen whether a diverse group of gatekeepers would bring in a more diverse group of scientific pioneers.

*What brings women to the ranks of scientific pioneers?* Like their male peers, women engage in science for both extrinsic and intrinsic rewards. Without exception, all subjects in this study reported the pleasures gained

from their activities. Aside from having a passion for science, they conducted
scientific investigations to receive recognitions and rewards. Political, economic, and social climates changed considerably since their childhoods and
adolescences. These changes were accompanied by personal efforts, commitment, and family as well as institutional support. All these circumstances
pushed subjects in this study in the direction of becoming a scientific pioneer.

Female scientific pioneers are made, not born, and the making of female
pioneers in science is anything but smooth. The process begins early in life
and continues into adulthood. This is consistent with Howe's thesis that scientific geniuses are the "product of a combination of environment, personality, and sheer hard work" (1999). When asked if they were born with certain
qualities, subjects in this study would probably say that being a pioneer in science can be learned. For the most part, these women survived and succeeded
in science because they believed in themselves. Despite upheavals in life and
work, they believed they had control over their surroundings. Others might
have given up under similar circumstances, but these women's outlook made
the decisive difference. They exhibited what Tausky and Dubin called the
"ambitious syndrome" (1965:729). These women were "upwardly anchored"
scientists. Holding an "unlimited success" view, they looked upward to maximum goals. This should not be surprising. These women further set themselves apart from and above their peers in terms of their exceptional ability to
handle both personal and professional challenges. The following comment on
Jonathan Cole's *Fair Science* (1979) spells out the requirements of survival
and success in science for women:

> This self-selected set of women, some of them superwomen and some who
> found superhusbands or marital situations, were able—surely by prodigious
> efforts—to push along as well as comparable men scientists despite domestic
> responsibilities (White 1982:955).

On the other hand, the experiences of subjects in the study do not meet the
assumption of gender symmetry within the family. According to Shauman
and Xie, women with human capital equal to those of their husbands usually
command a fully competitive role in making decisions regarding career
advancement (1996:458). This was apparently not the case for Marie
Curie, Maria Goeppert-Mayer, Dorothy Hodgkin, Rosalyn Yalow, and Fay
Ajzenberg-Selove. Whether married or single, these female scientific pioneers, as a group, were not highly geographically mobile due to historical and
personal circumstances.

In contrast, the less successful ones are "downwardly anchored" scientists
who tend to have a "limited success" view. However, it may be best for the

scientific community to have a mix of ambitious and less-than-ambitious people (Tausky and Dubin 1965:732). This argument holds if the career ambitions of individuals are not affected by functionally irrelevant characteristics.

*Who supports women to become scientific pioneers?* These female scientific pioneers managed to overcome the odds by themselves with the help of others. They had parents and teachers who motivated them. Family background and relationships shaped their career aspirations and career development. There were teachers and peers who wanted to help them achieve their maximum potential—to become high achievers. By parents and others, they were taught how to think positively and act realistically in light of difficult situations or negative outcomes. They learned to see and believe in new possibilities in bad times. Given the family and social upbringing of these female scientific pioneers, it is not surprising that they were indifferent to traditional gender-role socialization and other barriers for women entering science. They shared similar views about sources of discrimination. With the exception of Fay Ajzenberg-Selove, subjects in this study did not call for or work for collective changes in the practices and policies in science and society. For them, the best way to do better or the best way to overcome obstacles was "to be more qualified than men." Yalow's sign on the wall of her office at the Bronx VA Hospital—*Whatever women do they must do twice as well as men to be thought half as good* (Straus 1998:112)—endorses the common belief that one has to work twice as hard to get half as far. As mentioned in a previous section, their individual actions (a) actually undermined rather than promoted women's progress in science and (b) implied that women have themselves to blame for their disadvantaged status. Therefore, the career successes of these women might have actually hurt the prospects of other women in science. When large-scale and collaborative research becomes the norm in the scientific community, both collective measures as well as individual actions will be needed to improve women's participation in science.

*How do female scientists gain recognition?* It is impossible to make any broad statements about the success stories of these female scientific pioneers. All of them encountered a set of uncommon circumstances in their lives. Yet their ways of achieving success, in many respects, are no different from those used by men within and outside science. These female scientific pioneers clearly had political acumen, however. I can identify six things that helped them get to and stay at the top:

1. They understood that unwritten rules were as important as the official ones.
2. They understood the power structure in the scientific community.
3. They knew the "ropes" and built up connections to the world of science.

4. They were determined, but willing to compromise in dealings with sponsors or employers.
5. They listened to criticisms and took advice from others.
6. They kept working.

All this suggests that individuals who understand the unspoken (but important) "rules of the game" and follow them are more likely to succeed than those who do not (Gibbons, Holden, and Kaiser 1996:6). In this sense, these female pioneers were scientific entrepreneurs, and were the best rules-players in a variety of scientific environments.

Succeeding in science could resemble a tournament model, a social Darwinian argument proposed by Rosenbaum (1984). Yet a scientist's career is not a series of equal contests. The scientific community selects the "fittest." Aside from preexisting traits, structural and institutional support is associated with scientific career outcomes. Results of this study challenge the ideology of meritocracy, which encourages people to believe that individuals have access to similar opportunities to improve their relative standing in the society (Rosenbaum 1984:125). Subjects in this study did not buy into this ideology. In practice, they all realized the awesome power of netvigating as well as the operation of preferentialism and self-interestedness.

There is support for the social control of the production of female scientific pioneers. Some in this study had longstanding ties with the scientific community and some did not. However, all relied disproportionately on the old-boy network. They collaborated primarily with men (with the exception of Marie Curie, who worked closely with Irene Joliot-Curie after the death of Pierre Curie). Many obtained jobs the old-fashioned way—through networks. They were not deviants by conventional standards in science (Cottrell 1962). Due to the small number of women in science, men have traditionally dominated the roles of gatekeepers of resources and decision-makers. Women need and seek the support of men to make it in a man's world, a dilemma that female scientists generally face. Based on the actions of the scientific pioneers in this study, one can argue that they subscribed to the culture of science, and that they helped perpetuate the traditional scientific role model. First, they were intensely dedicated to achieving success in science. It was normal not to have outside interests. Second, rather than blaming or changing the structure of science, they adapted to the male culture of science. Third, they showed that women could make contributions to science within a male domain without changing the existing norms as the measure of excellence.

*What kind of impact will the actions and behavior of female scientific pioneers have on future generations of scientists?* Female pioneers-to-be will form a new class of elite scientists. Unlike traditional scientific elites, these female scientists will assume a wider range of roles in the scientific community. Not only will they succeed by doing well in and changing the game of science, but also by using the changes in society and science to set different directions for their careers. I argue that the scenario predicted by Chodorow is *unlikely* to happen: "As these eminent women retired, women's power and visibility in the field also declined. The leading women rose to the top and disappeared. They were not replaced, or, more actively, did not reproduce their own presence" (1991:172). Quite the contrary: many subjects in this study deliberately practiced professional generativity—passing on the craft to others (e.g., Curie, Joliot-Curie, Mead, Levi-Montalcini, Hodgkin, Yalow, and Ajzenberg-Selove). Junior scientists and students have sought female scientific pioneers out as role models or mentors. Many of these women have also been appointed to prestigious scientific societies or public offices. Like their male mentors or supporters, they have worn many hats throughout their careers. Further, the successful development of cross-gender networks among female scientific pioneers promotes integration within the scientific community.

Female scientists, including pioneers, still face the challenges of balancing family and work. Experiences of subjects in this study reveal a realistic possibility of balancing work and personal life among successful professional women. One can also make the argument that their experiences and testimonies reinforce the conventional norms in both the society and scientific community—that as women, they are expected to be nurturing and unselfish, and as scientists, they have to be competitive and strategic. As both rule-breakers and rule-makers, many subjects in this study met the expectations of these seemingly conflicting roles and were able to derive personal fulfillment and success from family and work.

*How do the experiences of women in science during the twentieth century represent or reflect the reality of recent experiences of women in science?* Scientific practices have changed appreciably in the past few decades; information on female scientific pioneers in the last century might be irrelevant to the current situation. To what extent can we generalize the findings of this study to more recent cohorts of female scientists? I cannot offer a definitive answer. It is difficult to make direct comparisons because of differences in preexisting circumstances. Having the right qualities and ideal circumstances does not necessarily produce female scientific pioneers. Yet nearly all the subjects in this study shared these characteristics or backgrounds. New generations of female scientists do not have to deal with formal barriers to education and employment. Many restrictive policies and practices are now gone. In the last few decades, schools and employers in the United States have been under pressure to show that they do not discriminate against women and members of other groups. The government, schools, and employers have implemented a range of programs and measures to improve women's participation in science. We may be able to observe the effects of these pressures on the careers of new generations of scientists (Long, Allison, and McGinnis 1993:720). So, compared to their predecessors, potential female pioneers may have fewer disadvantages in science. Another piece of good news for young female scientists is that women can have family and career and still be successful. But this should not be the benchmark for measuring personal and professional successes.

## CONCLUSION

This book provides a sociological analysis of female scientific pioneers, and contributes to an understanding of mobility in science by studying extreme cases (e.g., Blair-Loy 2003; Massey et al. 2003; O'Donovan-Polten 2001). Could we learn something different if a different sample of female scientific pioneers were used? Although the findings might be different from those presented in this book, the general patterns of career development and advancement would be similar. The strategies these women adopted may also be similar to those of men.

If it is difficult to be a female scientist, it is even more difficult to be a female scientific pioneer. To make it in science is similar to running a political campaign. Politicians have different styles and strategies, yet they all subscribe to certain rules and regulations. However, those who are highly successful in the political arenas or other fields tend to have distinct leadership skills (Gardner and Larkin 1995).

To what extent can we apply the lessons learned from female pioneers in science to women with outstanding careers in other sectors and vice versa? The production of successful female scientists may not be very different from the making of female leaders in business and industry. A combination of individual, structural, and institutional factors is required to make it in virtually any field or sector (e.g., Barker 1999; Chetkovich 1997; Harrington and Boardman 1997; Kunda 1992; O'Donovan-Polten 2001; Rosenbaum 1984; Zweigenhaft and Domhoff 1998).

This is the first sociological study of women who achieved tremendous success in a comparative perspective. I argue that there are many different ways in which women can be scientific pioneers. One could not extract an "ideal type" for success in science from ten exceptional female scientists. Contrary to expectations, there are *no* "best practices" among women succeeding in science. A closer look at the experiences of female scientific pioneers highlights the gap between *what we know* about female scientists and *what we should know* about female scientists. Analysis of existing biographical and ethnographical data reveals both universality and variations in background, circumstances, and strategies among the ten female scientific pioneers.

The notion of a golden rule to achieve success in science is a myth. There are both similarities and differences in styles, strategies, and career paths among female scientific pioneers. Ten women took different paths to achieve recognition and success, yet they all shared similar experiences while overcoming numerous obstacles. Regardless of their backgrounds and circumstances, subjects in this study, like the creative geniuses analyzed by Simonton, are "not privileged with any shortcuts to success" (1988:198). The activities and work habits of subjects in this study show that they do more than they have to. As a result, they get more than expected. The tenacity and stamina they all displayed indirectly bolsters Simonton's theory of chance permutations—that "geniuses are right more often only because they are wrong more often" (1988:191).

Aside from being rule-breakers and rule-makers, creating new roles for themselves (rather than conforming to traditional gender roles or simply breaking them) was their key to success. Setting new standards continuously, rather than meeting them, was *the* theme in their life and work.

The pattern of these female scientific pioneers was *not* to have patterns. Whether trying to balance work and family or in conducting research, few subjects in this study followed established gender or scientific roles. On the contrary, many of them helped change the status quo by rewriting some of the rules. There is variation in their style or approach to problems in life and work. Even so, there is no support for the argument that these subjects thought or engaged in science differently from men. The diverse experiences of the sample also demonstrate that one style or approach might not fit the demands of different situations.

This book does not provide an exhaustive treatment of female scientific pioneers. Only ten were included in the study. Instead, it provides insights into how a certain segment of female scientists took advantage of the opportunities in the twentieth century in pursuit of pioneering work. It also addresses a specific set of issues about women in science and offers a synthesis of explanations that tell us a good deal about their production from different fields, backgrounds, and cultures. Subjects in this study were a different breed of scientists, as the subtitle of the book might suggest.

Would these female scientists have been able to accomplish *much* more in the age of equality? Is there enough room for *all* who want to contribute to science? Because of the increasing importance of science to the society, there will always be opportunities in the scientific community waiting for those who want to be a part of it. The emergence of female scientific pioneers will not disappear. More female scientific pioneers are in the making, and this trend will continue for a pragmatic reason. When large-scale team research becomes the norm, female scientists will be more acceptable in the scientific community.

To reiterate a point made earlier in this chapter, female scientific pioneers are not born or self-made. They appear under certain circumstances and at certain times. A similar argument has been made by Blalock (1967) and Stephan and Levin (1992). My argument for a greater acceptance of women in science bears semblance to Mead's account of exclusiveness in fraternities, sororities, and professional occupations (Mead 1972:95–96).

There are more female scientific pioneers at work right now than at any other time in history. Numerous biographical anthologies concerning women in science have appeared in the last decade. Some of these works pay greater attention to the more recent generations of female scientists (e.g., Ambrose et al. 1977; Henrion 1997; Murray 2000; O'Connell 2001; Reynolds 1999; Wasserman 2000). Many women with outstanding careers who went into science in the years since the Second World War or in the years since Title IX was enacted in the United States are in fields such as genetics, mathematics, physics, psychology, and neuroscience.

In sum, more female scientists joining the ranks of pioneers is a direct, logical outcome of taking advantage of structural and institutional opportunities. As an emerging new class, female scientific pioneers could alter the distribution of power between gatekeepers and others.

# Notes

1. In a study of high school students' perceptions of barriers to women in science and engineering, one of the questions asked the respondents in 9th through 12th grades at a high school in Brooklyn, New York City (N = 212) was to name several male and female scientists. Most students, regardless of gender, had no difficulty naming three notable male scientists. But this was not the case when it came to providing the names of female scientists. The top five most popular scientists named by high school students were Albert Einstein, Marie Curie, Charles Darwin, Issac Newton, and Thomas Edison (Chan 2001). The same phenomenon was found among college students. In an Introduction to Sociology class (N = 104) offered by the author in fall 2002, college students also had no trouble with naming popular or famous male scientists. A list of three names of famous female scientists (Marie Curie, Rosalind Franklin, and Anna Freud) was generated, compared to a list of 20 names of famous male scientists (Arnold Armstrong, Francis Bacon, Alexander Graham Bell, Niels Bohr, the Wright Brothers (Wilbur and Orville), Francis Crick, Charles Darwin, Albert Einstein, Benjamin Franklin, Galileo Galilei, David Ho, William James, Gregor Mendel, Issac Newton, Ivan Pavlov, Ernest Rutherford, Carl Sagan, B.F. Skinner, Alfred Wallace, and James Watson).

2. It is important to note that many of Curie's papers have been restricted, so most of what we know about Marie Curie is based on the biography by her second daughter, Eve (Curie 1937). The public is familiar with her image of being a devoted wife and mother and exceptional researcher. Curie's capacity of establishing and running the Radium Institute, however, reveals that she was also one of the most enterprising scientists of the time, but this side of her has never been examined.

3. Margaret Mead relied more heavily on Ruth Benedict, with whom she even had an affair, than Franz Boas. She kept much of what was known about Benedict and herself restricted for years (Banner 2003; Lapsley 1999).

4. Although it is beyond the scope of this work, it is important to note that Margaret Mead differed from other subjects in the sample of this study in terms of her

marital instability and her unconventional attitudes toward sexuality (Banner 2003; Lapsley 1999).

5. It is beyond the scope of this work to examine Rachel Carson's life at great length. After the publication of the Rachel Carson and Dorothy Freeman letters, much more has been written about Carson's personal life. The release of these letters gave rise to an extensive discussion of Carson's sexuality and its impact on her work (Freeman 1995).

6. Although many of the subjects in this study are immigrants, the focus of this study is on the United States. It is imperative to note that conditions for female scientists vary widely across regions such as the former Soviet Union, the Middle East, Scandinavia, England, Italy, Australia, as well as South American and African countries. For this reason, the experiences of the women in this study might not be applicable to women in other places and times.

7. The discussions of alternative career models in this section in no way suggest that married women with children are the most significant role models for aspiring women in science. They do not promote the viewpoint that valorizes heterosexual marriage and traditional gender roles.

8. In the preface of her biography of Barbara McClintock, Keller noted her subject's initial resistance to being interviewed, because McClintock "said she was too much of a maverick for her story to be of interest to others. . . . But as the interviews progressed, [Keller] drew confirmation for [her] instinct from [McClintock's] own emphasis on *how much can be learned from the exceptional, 'anomalous' example*" [emphasis added] (1983: xxii).

9. The purpose of this section is not to impose a value judgment on geniuses and other scientists. The discussions should not be construed as an acceptance of the value hierarchy of elite science, at least as it is conducted in the United States and Europe. The reference to Albert Einstein and Marie Curie in the section heading does not necessarily mean they are representative role models for twentieth century science. The everyday conduct of "normal science" is not less important than scientific breakthroughs.

# Bibliography

Abir-Am, Pnina G. and Dorinda Outram, eds. 1987. *Uneasy Careers and Intimate Lives: Women in Science, 1789–1979*. New Brunswick, NJ: Rutgers University Press.

Ainley, Marianne Gosztonyi, ed. 1990. *Despite the Odds: Essays on Canadian Women and Science*. Montreal: Véhicule Press.

Ajzenberg-Selove, Fay. 1994. *A Matter of Choices: Memoirs of a Female Physicist*. New Brunswick, NJ: Rutgers University Press.

Allison, Paul D. 1980. *Processes of Stratification*. New York: Arno Press.

Allison, Paul D. 1992. "The Cultural Evolution of Beneficent Norms." *Social Forces* 71(2):279–301.

Ambrose, Susan A., Kristin L. Dunkle, Barbara B. Lazarus, Indira Nair, and Deborah A. Harkus. 1997. *Journeys of Women in Science and Engineering: No Universal Constants*. Philadelphia: Temple University Press.

Banner, Lois W. 2003. *Intertwined Lives: Margaret Mead, Ruth Benedict, and Their Circle*. New York: Knopf.

Barber, Leslie A. 1995. "U.S. Women in Science and Engineering, 1960–1990: Progress Toward Equity?" *Journal of Higher Education* 66(2):213–234.

Barker, Joan C. 1999. *Danger, Duty, and Disillusion: The Worldview of Los Angeles Police Officers*. Prospect Heights, IL: Waveland Press.

Bengtson, Vern L., Timothy J. Biblarz, and Robert E.L. Roberts. 2002. *How Families Still Matter: A Longitudinal Study of Youth in Two Generations*. New York: Cambridge University Press.

Bensaude-Vincent, Bernadette. 1996. "Star Scientists in a Nobelist Family: Irene and Frederic Joliot-Curie." Pp. 57–71 in *Creative Couples in the Sciences*, edited by Helena M. Pycior, Nancy G. Slack, and Onina G. Abir-Am. New Brunswick, NJ: Rutgers University Press.

Berg, Ivar and Arne L. Kalleberg, eds. 2001. *Sourcebook of Labor Markets: Evolving Structures and Processes*. New York: Plenum.

Bernhard, Carl Gustaf. 1997. "The Nobel Prizes and Nobel Institutions." Pp. 37–42 in *Nobel Prize Winners: 1992–1996 Supplement*, edited by Clifford Thompson. New York: H.W. Wilson Company.

Berry, Colin. 1981. "The Nobel Scientists and the Origins of Scientific Achievement." *British Journal of Sociology* 32(3):381–391.

Blair-Loy, Mary. 2003. *Competing Devotions: Career and Family among Women Executives*. Cambridge, MA: Harvard University Press.

Blalock, Hubert M., Jr. 1967. *Toward a Theory of Minority-Group Relations*. New York: Capricorn Books.

Booth, William. 1989. "Oh, I Thought You Were a Man." *Science* (January 27):475.

Bradford, Janet. 1980. "Women Scientists in New Zealand—Why So Few?" *Impact of Science on Society* 30(1):37–42.

Carson, Rachel. 1951. *The Sea Around Us*. New York: Oxford University Press.

Carson, Rachel. 1962. *Silent Spring*. Boston: Houghton Mifflin.

Chakravarthy, R., A. Chawla, and G. Mehta. 1988. "Women Scientists at Work—An International Comparative Study of Six Countries." *Scientometrics* 14(1&2):43–74.

Chan, Susanna. 2001. "High School Students' Perceptions of Barriers to Women in U.S. Science and Engineering." A Social Science Project for the *2001–02 Intel Science Talent Search*, Midwood High School at Brooklyn College of the City University of New York.

Chetkovich, Carol. 1997. *Real Heat: Gender and Race in the Urban Fire Service*. New Brunswick, NJ: Rutgers University Press.

Chodorow, Nancy J. 1991. "Where Have All the Eminent Women Psychoanalysts Gone? Like the Bubbles in Champagne, They Rose to the Top and Disappeared." Pp. 167–194 in *Social Roles and Social Institutions: Essays in Honor of Rose Laub Coser*, edited by Judith R. Blau and Norman Goodman. Boulder, CO: Westview Press.

*Chronicle of Higher Education*. 2004. "Canada's Billion Dollar Controversy: A Major Attempt to Attract Research Stars that Netted Few Women, Leading to Charges of Bias." Vol. L, No. 18 (January 19):A38–A39.

Clark, Roger D. 1982. "Birth Order and Eminence: A Study of Elites in Science, Literature, Sports, Acting and Business." *International Review of Modern Sociology* 12(Autumn):273–289.

Clark, Roger D. and Glenn A. Rice. 1982. "Family Constellations and Eminence: The Birth Orders of Nobel Prize Winners." *Journal of Psychology* 110:281–287.

Cole, Jonathan R. 1979. *Fair Science: Women in the Scientific Community*. New York: Free Press.

Cole, Stephen. 1992. *Making Science: Between Nature and Society*. Cambridge, MA: Harvard University Press.

Comfort, Nathaniel C. 2001. *The Tangled Field: Barbara McClintock's Search for the Patterns of Genetic Control*. Cambridge, MA: Harvard University Press.

Cottrell, A.H. 1962. "Scientists: Solo or Concerted?" Pp. 388–393 in *The Sociology of Science*, edited by Bernard Barber and Walter Hirsch. New York: Free Press.

Crane, Diana. 1965. "Scientists at Major and Minor Universities: A Study of Productivity and Recognition." *American Sociological Review* 30(5):699–714.

Crane, Diana. 1967. "The Gatekeepers of Science: Some Factors Affecting the Selection of Articles for Scientific Journals." *American Sociologist* 2:195–201.

Crane, Diana. 1972. *Invisible Colleges: Diffusion of Knowledge in Scientific Communities*. Chicago: University of Chicago Press.

Crawford, Elizabeth. 1998. "Nobel: Always the Winners, Never the Losers." *Science* 282(November 13):1256–1257.

Creese, Mary R.S. 1998. *Ladies in the Laboratory? American and British Women in Science, 1800–1900: A Survey of Their Contributions to Research*. Lanham, MD: Scarecrow Press.

Crossfield, E. Tina. 1997. "Irene Joliot-Curie: Following in Her Mother's Footsteps." Pp. 97–123 in *A Devotion to Their Science: Pioneer Women of Radioactivity*, edited by M.F. Rayner-Canham and G.W. Rayner-Canham. Philadelphia: Chemical Heritage Foundation and Montreal: McGill-Queen's University Press.

Curie, Eve. 1937. *Madame Curie*. New York: Doubleday.

Curie, Marie 1936. *Pierre Curie*. New York: Macmillan.

Dash, Joan. 1973. "Maria Goeppert-Mayer." Pp. 229–346 in *A Life of One's Own: Three Gifted Women and the Men They Married*. New York: Harper & Row.

de Meuron-Landolt, Monique. 1975. "How a Woman Scientist Deals Professionally with Men." *Impact of Science on Society* 25(2):147–152.

Driscoll, Dawn-Marie and Carol R. Goldberg. 1993. *Members of the Club: The Coming of Age of Executive Women*. New York: Free Press.

Eisenhart, Margaret and Elizabeth Finkel. 1998. *Women's Science: Learning and Succeeding from the Margins*. Chicago: University of Chicago Press.

Epstein, Cynthia Fuchs. 1970. "Encountering the Male Establishment: Sex-Status Limits on Women's Careers in the Professions." *American Journal of Sociology* 75:965–982.

Epstein, Cynthia Fuchs. 1993. *Women in Law*. 2d ed. Urbana: University of Illinois Press.

Etzkowitz, Henry, Carol Kemelgor, and Brian Uzzi. 2000. *Athena Unbound: The Advancement of Women in Science and Technology*. New York: Cambridge University Press.

Evetts, Julia. 1996. *Gender and Career in Science and Engineering*. Bristol, PA: Taylor & Francis.

Fausto-Sterling, Anne. 1992. *Myths of Gender: Biological Theories About Women and Men*. 2d ed. New York: Basic Books.

Fedoroff, Nina V. 1996. "Two Women Geneticists." *American Scholar* 65(4):587–592.

Ferry, Georgina. 1998. *Dorothy Hodgkin: A Life*. Cold Spring Harbor, NY: Cold Spring Harbor Laboratory Press.

Fox, Mary Frank and Paula E. Stephan. 2001. "Careers of Young Scientists: Preferences, Prospects and Realities by Gender and Field." *Social Studies of Science* 31(1):109–122.

Freeman, Martha, ed. 1995. *Always, Rachel*. Boston: Beacon.

Gardner, Howard and Emma Laskin. 1995. *Leading Minds: An Anatomy of Leadership*. New York: Basic Books.

Gibbons, Ann, Constance Holden, and Jocelyn Kaiser. 1996. "Facing the Big Chill in Science." *Science* 271(5257):1902–1905.

Gieryn, Thomas F. 1995. "Boundaries of Science." Pp. 393–443 in *Handbook of Science and Technology Studies*, edited by Sheila Jasanoff, Gerald E. Markle, James C. Petersen, and Trevor Pinch. Thousand Oaks, CA: Sage.

Glover, Judith. 2000. *Women and Scientific Employment*. New York: St. Martin's Press.

Goertzel, Mildred George, Victor Goertzel, and Ted George Goertzel. 1978. *Three Hundred Eminent Personalities: A Psychosocial Analysis of the Famous*. San Francisco: Jossey-Bass Publishers.

Goertzel, Victor and Mildred George Goertzel. 1962. *Cradles of Eminence*. Boston: Little, Brown and Company.

Granovetter, Mark. 1995. *Getting a Job: A Study of Contacts and Careers*. 2d. Chicago: University of Chicago Press.

Gray, George W. 1962. "Which Scientists Win Nobel Prizes?" Pp. 557–565 in *The Sociology of Science*, edited by Bernard Barber and Walter Hirsch. New York: Free Press.

Greenstein, George. 1998. *Portraits of Discovery: Profiles in Scientific Genius*. New York: John Wiley & Sons.

Grinager, Patricia. 1999. *Uncommon Lives: My Lifelong Friendship with Margaret Mead*. Lanham, MD: Rowman & Littlefield.

Gustin, Bernard H. 1973. "Charisma, Recognition, and the Motivation of Scientists." *American Journal of Sociology* 78(5):1119–1134.

Hagstrom, Warren O. 1974. "Competition in Science." *American Sociological Review* 39(1):1–18.

Hanson, Sandra L. 1996. *Lost Talent: Women in the Sciences*. Philadelphia: Temple University Press.

Hargens, Lowell L. 1978. "Relations Between Work Habits, Research Technologies, and Eminence in Science." *Sociology of Work and Occupations* 5(1):97–112.

Harrington, Charles C. and Susan K. Boardman. 1997. *Paths to Success: Beating the Odds in American Society*. Cambridge, MA: Harvard University Press.

Helson, Ravenna. 1971. "Women Mathematicians and the Creative Personality." *Journal of Counseling and Clinical Psychology* 36(2):210–220.

Helson, Ravenna. 1996. "Arnheim Award Address to Division 10 of the American Psychological Association." *Creativity Research Journal* 9(4):295–306.

Henrion, Claudia. 1997. *Women in Mathematics: The Addition of Difference*. Bloomington, Indiana: Indiana University Press.

Hermanowicz, Joseph C. 1998. *The Stars Are Not Enough: Scientists—Their Passions and Professions*. Chicago: University of Chicago Press.

Hodson, Randy. 2001. *Dignity at Work*. New York: Cambridge University Press.

Howe, Michael J.A. 1999. *Genius Explained*. New York: Cambridge University Press.

Inhaber, H. and K. Przednowek. 1976. "Quality of Research and the Nobel Prizes." *Social Studies of Science* 6:33–50.

Jacobs, Jerry A. 1989. *Revolving Doors: Sex, Segregation and Women's Careers*. Stanford, CA: Stanford University Press.

Jacobs, Jerry A., ed. 1995. *Gender Inequality at Work*. Thousand Oaks, CA: Sage.

Jones, L.M. 1990. "Intellectual Contributions of Women to Physics." Pp. 188–214 in *Women of Science: Righting the Record*, edited by G. Kass-Simon and Patricia Farnes. Bloomington: Indiana University Press.

Julian, Maureen M. 1990. "Women in Crystallography." Pp. 335–383 in *Women of Science: Righting the Record*, edited by G. Kass-Simon and Patricia Farnes. Bloomington: Indiana University Press.

Kalleberg, Arne L. and Aage B. Sorensen. 1979. "The Sociology of Labor Markets." *Annual Review of Sociology* 5:352–379.

Kanter, Rosabeth Moss. 1977. *Men and Women of the Corporation*. New York: Basic Books.

Keller, Evelyn Fox. 1983. *A Feeling for the Organism: The Life and Work of Barbara McClintock*. New York: W.H. Freeman & Company.

Keller, Evelyn Fox. 1985. *Reflections on Gender and Science*. New Haven, CT: Yale University Press.

Kerckhoff, Alan C., ed. 1996. *Generating Social Stratification: Toward a New Research Agenda*. Boulder, CO: Westview Press.

Kohlstedt, Sally Gregory, ed. 1999. *History of Women in the Sciences: Readings from ISIS*. Chicago: University of Chicago Press.

Kuhn, Thomas S. 1962. *The Structure of Scientific Revolutions*. Chicago: University of Chicago Press.

Kulis, Stephen and Karen A. Miller. 1988. "Are Minority Women Sociologists in Double Jeopardy?" *The American Sociologist* 19(4):323–339.

Kunda, Gideon. 1992. *Engineering Culture: Control and Commitment in a High-Tech Corporation*. Philadelphia: Temple University Press.

LaFollette, Marcel C. 1988. "Eyes on the Stars: Images of Women Scientists in Popular Magazines." *Science, Technology, & Human Values* 13(3&4):262–275.

Lapsley, Hilary. 1999. *Margaret Mead and Ruth Benedict: The Kinship of Women*. Amherst: University of Massachusetts.

Lear, Linda. 1998. *Rachel Carson: Witness for Nature*. New York: Henry Holt and Company.

Levi-Montalcini, Rita. 1988. *In Praise of Imperfection: My Life and Work*. New York: Basic Books.

Lipman-Blumen, Jean. 1976. "Toward a Homosocial Theory of Sex Roles: An Explanation of the Sex Segregation of Social Institutions." Pp. 15–31 in *Women and the Workplace: The Implications of Occupational Segregation*, edited by Martha Blaxall and Barbara B. Reagan. Chicago: University of Chicago Press.

Liversidge, Anthony. 1988. "Interview: Rita Levi-Montalcini." *Omni* 10(6):70–74, 102–105.

Long, J. Scott. 1992. "Measures of Sex Differences in Scientific Productivity." *Social Forces* 71(1):159–178.

Long, J. Scott, Paul D. Allison, and Robert McGinnis. 1979. "Entrance into the Academic Career." *American Sociological Review* 44(5):816–830.

Long, J. Scott, Paul D. Allison, and Robert McGinnis. 1993. "Sex Differences and Productivity in Academic Rank Advancement." *American Sociological Review* 58(5):703–722.

Long, J. Scott and Mary Frank Fox. 1995. "Scientific Careers: Universalism and Particularism." *Annual Review of Sociology* 21:45–71.

Lonsdale, Kathleen. 1970. "Women in Science: Reminiscences and Reflections." *Impact of Science on Society* 20(1):45–59.

Maccoby, Eleanor E. 1970. "Feminine Intellect and the Demands of Science." *Impact of Science on Society* 20(1):13–28.

Margolis, Jane and Allan Fisher. 2002. *Unlocking the Clubhouse: Women in Computing*. Cambridge, MA: MIT Press.

Mark, Joan. 1999. *Margaret Mead: Coming of Age in America*. New York: Oxford University Press.

Massey, Douglas S., Camille Z. Charles, Garvey F. Lundy, and Mary J. Fischer. 2003. *The Source of the River: The Social Origins of Freshmen at America's Selective Colleges and Universities*. Princeton: Princeton University Press.

McCay, Mary A. 1993. *Rachel Carson*. New York: Twayne Publishers.

McGrayne, Sharon Bertsch. 1998. *Nobel Prize Women in Science: Their Lives, Struggles, and Momentous Discoveries*. 2d ed. Secaucus, NJ: Carol Publishing Group.

McIlwee, Judith S. and J. Gregg Robinson. 1992. *Women in Engineering: Gender, Power, and Workplace Culture*. Albany: State University of New York Press.

McKown, Robin. 1961. *She Lived for Science: Irene Joliot-Curie*. New York: Julian Messner, Inc.

Mead, Margaret. 1928. *Coming of Age in Samoa: A Psychological Study of Primitive Youth for Western Civilization*. New York: William Morrow.

Mead, Margaret. 1972. *Blackberry Winter: My Earlier Years*. New York: William Morrow.

Merton, Robert K. 1973. *The Sociology of Science: Theoretical and Empirical Investigations*. Chicago: University of Chicago Press.

Michalko, Michael. 1998. *Cracking Creativity: The Secrets of Creative Genius*. Berkeley, CA: Ten Speed Press.

Mitroff, Ian I. 1974. *The Subjective Side of Science: A Philosophical Inquiry into the Psychology of the Apollo Moon Scientists*. New York: Elsevier.

Modis, Theodore. 1988. "Competition and Forecasts for Nobel Prize Awards." *Technological Forecasting and Social Change* 34:95–102.

Monroe, Kristen Renwick. 1996. *The Heart of Altruism: Perceptions of a Common Humanity*. Princeton: Princeton University Press.

Morgan, Carolyn Stout. 1992. "College Students' Perceptions of Barriers to Women in Science and Engineering." *Youth & Society* 24(2):228–236.

Mortimer, Jeylan T. 2003. *Working and Growing Up in America*. Cambridge, MA: Harvard University Press.

Moss, Philip and Chris Tilly. 2001. *Stories Employers Tell: Race, Skill, and Hiring in America*. New York: Russell Sage Foundation.

Moulin, Leo. 1955. "The Nobel Prizes for the Sciences from 1901–1950—An Essay in Sociological Analysis." *British Journal of Sociology* 6(3):246–263.

Mozans, H.J. 1991. *Woman in Science*. Notre Dame, IN: University of Notre Dame Press.

Murray, Margaret A.M. 2000. *Women Becoming Mathematicians: Creating a Professional Identity in Post-World War II America*. Cambridge, MA: MIT Press.

National Science Foundation. 1994. *Women, Minorities, and Persons with Disabilities in Science and Engineering*. Arlington, VA: National Science Foundation (NSF 94–333).

National Science Foundation. 1999. *Women, Minorities, and Persons with Disabilities in Science and Engineering*. Arlington, VA: National Science Foundation (NSF 99–338).

*New Scientist*. 1994. "Behind the Nobel Glitz." (December 14):3.

Nidiffer, Jana. 2000. *Pioneering Deans of Women: More than Wise and Pious Matrons*. New York: Teachers College Press.

Nobel Foundation. 2004. "Board of Directors." (http://nobelprize.org/nobel/nobel-foundation/directors.html).

Nowotny, Helga. 1991. "Mixed Feelings: Women Interacting with the Institution of Science." Pp. 149–165 in *Social Roles and Social Institutions: Essays in Honor of Rose Laub Coser*, edited by Judith R. Blau and Norman Goodman. Boulder, CO: Westview Press.

Ochse, R. 1990. *Before the Gates of Excellence: The Determinants of Creative Genius*. New York: Cambridge University Press.

Ochse, Rhona. 1991. "Why Were There Relatively Few Eminent Women Creators?" *Journal of Creative Behavior* 25(4):334–343.

O'Connell, Agnes N., ed. 2001. *Models of Achievement: Reflections of Eminent Women in Psychology*, Vol. 3. Mahwah, NJ: Lawrence Erlbaum Associates.

O'Donovan-Polten, Sheelagh. 2001. *The Scales of Success: Constructions of Life-Career Success of Eminent Men and Women Lawyers*. Toronto: University of Toronto Press.

Ogilvie, Marilyn and Joy Harvey, eds. 2000. *The Biographical Dictionary of Women in Science: Pioneering Lives from Ancient Times to the Mid-20th Century*. New York: Routledge.

Oliner, Samuel P. and Pearl M. Oliner. 1988. *The Altruistic Personality: Rescuers of Jews in Nazi Europe*. New York: Free Press.

Ortner, Sherry B. 1999. *Life and Death on Mt. Everest: Sherpas and Himalayan Mountaineering*. Princeton: Princeton University Press.

Osterman, Paul. Ed. 1996. *Broken Ladders: Managerial Careers in the New Economy*. New York: Oxford University Press.

Ostrower, Francie. 1995. *Why the Wealthy Give: The Culture of Elite Philanthropy*. Princeton: Princeton University Press.

Ostrower, Francie. 2002. *Trustees of Culture: Power, Wealth, and Status on Elite Arts Boards*. Chicago: University of Chicago Press.

Pasachoff, Naomi. 1996. *Marie Curie and the Science of Radioactivity*. New York: Oxford University Press.

Pattatucci, Angela M., ed. 1998. *Women in Science: Meeting Career Challenges*. Thousand Oaks, CA: Sage.

Polkinghorne, John. 1998. *Beyond Science: The Wider Human Context*. New York: Cambridge University Press.

Powell, Gary N. and Laura M. Graves. 2003. *Women and Men in Management*. 3rd ed. Thousand Oaks, CA: Sage.

Preston, Anne E. 1994. "Why Have All the Women Gone? A Study of Exit of Women from the Science and Engineering Professions." *American Economic Review* 84:1446–1462.

Preston, Anne E. 2004. *Leaving Science: Occupational Exit from Scientific Careers*. New York: Russell Sage Foundation.

Pycior, Helena M. 1993. "Reaping the Benefits of Collaboration while Avoiding Its Pitfalls: Marie Curie's Rise to Scientific Prominence." *Social Studies of Science* 23:301–323.

Pycior, Helena M. 1996. "Pierre Curie & 'His Eminent Collaborator Mme. Curie': Complementary Partners." Pp. 39–56 in *Creative Couples in the Sciences*, edited by Helena M. Pycior, Nancy G. Slack, and Pnina G. Abir-Am. New Brunswick, NJ: Rutgers University Press.

Pycior, Helena M., Nancy G. Slack, and Pnina G. Abir-Am, eds. 1996. *Creative Couples in the Sciences*. New Brunswick, NJ: Rutgers University Press.

Quinn, Susan. 1995. *Marie Curie: A Life*. Cambridge, MA: Perseus Books.

Ramon y Cajal, Santiago. 1999. *Advice for a Young Investigator*. Cambridge, MA: MIT Press.

Rayner-Canham, Marelene F. and Geoffrey W. Rayner-Canham, eds. 1997. *A Devotion to Their Science: Pioneer Women in Radioactivity*. Philadelphia and Montreal: Chemical Heritage Foundation and McGill-Queen's University Press.

Rayner-Canham, Marelene F. and Geoffrey W. Rayner-Canham. 1998. *Women in Chemistry: Their Changing Roles from Alchemical Times to the Mid-Twentieth Century*. Philadelphia: American Chemical Society and Chemical Heritage Foundation.

Reid, Robert. 1974. *Marie Curie*. New York: Saturday Review Press/E.P. Dutton.

Reskin, Barbara F. 1976. "Sex Differences in Status Attainment in Science: The Case of Postdoctoral Fellowship." *American Sociological Review* 41:597–612.

Reskin, Barbara F. 1978. "Sex Differentiation and the Social Organization of Science." Pp. 6–37 in *Sociology of Science*, edited by Jerry Gaston. San Francisco: Jossey-Bass.

Reskin, Barbara F. 1979. "Academic Sponsorship and Scientists' Careers." *Sociology of Education* 52:129–146.

Reskin, Barbara F. 1998. *The Realities of Affirmative Action in Employment*. Washington, DC: American Sociological Association.

Reskin, Barbara F. and Patricia A. Roos. 1990. *Job Queues, Gender Queues: Explaining Women's Inroads into Male Occupations*. Philadelphia: Temple University Press.

Reynolds, Moira Davison. 1999. *American Women Scientists: 23 Inspiring Biographies, 1900–2000*. Jefferson, NC: McFarland & Company.

Richardson, Darlene S. and Connie J. Sutton. 1993. "Ordinary and Extraordinary Women in Science." *Bulletin of Science, Technology & Society* 13:251–254.

Roe, Anne. 1952. *The Making of a Scientist*. New York: Dodd, Mead, & Company.

Rose, Suzanna. 1989. "Women Biologists and the 'Old Boy' Network." *Women's Studies International Forum* 12(3):349–354.

Rosenbaum, James E. 1984. *Career Mobility in a Corporate Hierarchy*. New York: Academic Press.

Rosser, Sue V. 2000. *Women, Science, and Society: The Crucial Union*. New York: Teachers College Press.

Rosser, Sue V. 2004. *The Science Glass Ceiling: Academic Women Scientists and the Struggle to Succeed.* New York: Routledge.

Rossiter, Margaret W. 1982. *Women Scientists in America: Struggles and Strategies to 1940.* Baltimore, MD: Johns Hopkins University Press.

Rossiter, Margaret W. 1993. "The Matilda Effect in Science." *Social Studies of Science* 23:325–341.

Rossiter, Margaret W. 1995. *Women Scientists in America: Before Affirmative Action, 1940–1972.* Baltimore, MD: Johns Hopkins University Press.

Rossiter, Margaret W. 2003. "A Twisted Tale: Women in the Physical Sciences in the Nineteenth and Twentieth Centuries." Pp. 54–71 in *The Cambridge History of Science: The Modern Physical and Mathematical Sciences*, edited by Mary Jo Nye. Vol. 5. New York: Cambridge University Press.

Schiebinger, Londa. 1987. "The History and Philosophy of Women in Science: A Review Essay." *Signs: Journal of Women in Culture and Society* 12(2):305–332.

Schiebinger, Londa. 2001. *Has Feminism Changed Science?* Cambridge, MA: Harvard University Press.

Seymour, Elaine and Nancy M. Hewitt. 1997. *Talking About Leaving: Why Undergraduates Leave the Sciences.* Boulder, CO: Westview Press.

Shapley, Deborah. 1975. "Obstacles to Women in Science." *Impact of Science on Society* 25(2):115–123.

Shauman, Kimberlee A. and Yu Xie. 1996. "Geographic Mobility of Scientists: Sex Differences and Family Constraints." *Demography* 33(4):455–468.

Simmons, John. 1996. *The Scientific 100: A Ranking of the Most Influential Scientists, Past and Present.* Secaucus, NJ: Carol Publishing.

Simon, Rita James. 1974. "The Work Habits of Eminent Scholars." *Sociology of Work and Occupations* 1(3):327–335.

Simonton, Dean Keith. 1988. *Scientific Genius: A Psychology of Science.* Cambridge, MA: Cambridge University Press.

Simonton, Dean Keith. 1999. *Origins of Genius: Darwinian Perspectives on Creativity.* New York: Oxford University Press.

Sonnert, Gerhard and Gerald Holton. 1995. *Who Succeeds in Science? The Gender Dimension.* New Brunswick, NJ: Rutgers University Press.

Stanley, Autumn. 1995. *Mothers and Daughters of Invention: Notes for a Revised History of Technology.* New Brunswick, NJ: Rutgers University Press.

Stephan, Paula E. and Sharon G. Levin. 1992. *Striking the Mother Lode in Science: The Importance of Age, Place, and Time.* New York: Oxford University Press.

Stephan, Paula E. and Sharon G. Levin. 1993. "Age and the Nobel Prize Revisited." *Scientometrics* 28(3):387–399.

Stolte-Heiskanen, Veronica. 1983. "The Role and Status of Women Scientific Research Workers in Research Groups." *Research in the Interweave of Social Roles: Jobs and Families* 3:59–87.

Straus, Eugene. 1998. *Rosalyn Yalow: Nobel Laureate: Her Life and Work in Medicine.* New York: Plenum.

Sulloway, Frank J. 1996. *Born to Rebel: Birth Order, Family Dynamics, and Creative Lives.* New York: Vintage Books.

Swerdlow, Marian. 1998. *Underground Women: My Four Years as a New York City Subway Conductor*. Philadelphia: Temple University Press.

Tang, Joyce. 2003. "Women Succeeding in Science in the Twentieth Century." *Sociological Forum* 18(2):325–342.

Tang, Joyce. 2005. "Manufacturing Great Scientists." *Sociological Inquiry* 75(1):129–150.

Tausky, Curt and Robert Dubin. 1965. "Career Anchorage: Managerial Mobility Motivations." *American Sociological Review* 30(5):725–735.

*TIME 100: The Century's Greatest Minds—Scientists and Thinkers of the 20th Century*. 1999. (March 29), vol. 153, no. 2. New York: Time, Inc.

Tomaskovic-Devey, Donald. 1993. *Gender and Racial Inequality at Work: The Sources and Consequences of Job Segregation*. Ithaca, NY: ILR Press.

Tomaskovic-Devey, Donald, Arne L. Kalleberg, and Peter V. Marsden. 1996. "Organizational Patterns of Gender Segregation." Pp. 276–301 in *Organizations in America: Analyzing Their Structures and Human Resource Practices*, edited by Arne L. Kalleberg, David Knoke, Peter V. Marsden, and Joe L. Spaeth. Thousand Oaks, CA: Sage.

Tosi, Lucia. 1975. "Women's Scientific Creativity." *Impact of Science on Society* 25(2):105–114.

U.S. Department of Education. 1997. *Title IX: 25 Years of Progress*. (http://www.ed .gov/pubs/TitleIX/index.html).

Valian, Virginia. 1998. *Why So Slow? The Advancement of Women*. Cambridge, MA: MIT Press.

Wajcman, Judy. 1995. "Feminist Theories of Technology." Pp. 189–204 in *Handbook of Science and Technology Studies*, edited by Sheila Jasanoff, Gerald E. Markle, James C. Petersen, and Trevor Pinch. Thousand Oaks, CA: Sage.

Wasserman, Elga. 2000. *The Door in the Dream: Conversations with Eminent Women in Science*. Washington, DC: Joseph Henry Press.

*Webster's Third New International Dictionary*. 1981. Springfield, MA: Merriam-Webster, Inc.

White, Harrison C. 1982. "Review Essay: Fair Science?" *American Journal of Sociology* 87(4):951–956.

Wiegand, Sylvia. 1996. "Grace Chisholm Young and William Henry Young." Pp. 126–140 in *Creative Couples in the Sciences*, edited by Helena M. Pycior, Nancy G. Slack, and Pnina G. Abir-Am. New Brunswick, NJ: Rutgers University Press.

Williams, Christine L. 1989. *Gender Differences at Work: Women and Men in Nontraditional Occupations*. Berkeley: University of California Press.

Williams, Christine L. 1995. *Still a Man's World: Men Who Do Women's Work*. Berkeley: University of California Press.

Wolpert, Lewis and Alison Richards. 1988. *A Passion of Science*. New York: Oxford University Press.

Wright, Rosemary. 1997. *Women Computer Professionals: Progress and Resistance*. Lewiston, NY: Edwin Mellen Press.

Wuthnow, Robert. 1991. *Acts of Compassion: Caring for Others and Helping Ourselves*. Princeton: Princeton University Press.

Xie, Yu and Kimberlee A. Shauman. 2003. *Women in Science: Career Processes and Outcomes*. Cambridge, MA: Harvard University Press.

Yentsch, Clarice M. and Carl J. Sindermann. 1992. *The Woman Scientist: Meeting the Challenges for a Successful Career*. New York: Plenum.

Yoder, Janice D. 1994. "Looking Beyond Numbers: The Effects of Gender Status, Job Prestige, and Occupational Gender-Typing on Tokenism Processes." *Social Psychology Quarterly* 57(2):150–159.

Zimmer, Lynn. 1988. "Tokenism and Women in the Workplace: The Limits of Gender-Neutral Theory." *Social Problems* 35(1):64–77.

Zuckerman, Harriet. 1978. "The Sociology of the Nobel Prize: Further Notes and Queries—How Successful Are the Prizes in Recognizing Scientific Excellence?" *American Scientist* 66:420–425.

Zuckerman, Harriet. 1988. "The Sociology of Science." Pp. 511–574 in *Handbook of Sociology*, edited by Neil J. Smelser. Newbury Park, CA: Sage.

Zuckerman, Harriet. 1996. *Scientific Elites: Nobel Laureates in the United States*. New Brunswick, NJ: Transaction Publishers.

Zuckerman, Harriet, Jonathan R. Cole, and John T. Bruer. Eds. 1991. *The Outer Circle: Women in the Scientific Community*. New York: W.W. Norton.

Zweigenhaft, Richard L. and G. William Domhoff. 1998. *Diversity in the Power Elite: Have Women and Minorities Reached the Top?* New Haven, CT: Yale University Press.

# Index

Academy of the Sciences:
French, 44, 45, 107;
U.S., 50, 114
access to resources, 109;
gender differences in, 39
accumulative advantages, 34, 128.
*See also* cumulative advantages
accumulative disadvantages, 128.
*See also* cumulative disadvantages
affirmative action policies, 26, 30, 82, 121
Africa, 152n6
Ajzenberg-Selove, Fay, 9, 52, 64–65, 72, 74, 79, 82–83, 86, 90, 94, 96, 121–22, 127, 139, 142, 144–45, 147
altruism, 122–23
ambitious syndrome, 144
American Anthropological Association, 142
American Association for the Advancement of Science (AAAS), 142
*American Journal of Sociology*, 104
American Museum of Natural History, 46, 96
American Physical Society, 142
Amprino, Rodolfo, 117
Angeletti, Pietro, 56, 118

anthropology, 99;
physical, 112
antifeminist, 57–58
antinepotism, 52, 78, 107
anti-Semitism, 56
apprenticeship, 29, 109
approach:
biological, xi, 22–24, weaknesses of, 25;
collaborative, 47;
individual choice, xi, 24–27, 40, weaknesses of, 25;
institutional, xi;
life-course, 3–4, 17, 40;
networking, 129;
structural, xi;
transdisciplinary, 47
approach to:
problem solving, 12–13, 91;
scientific investigations, 2–3, 12, 60
archaeology, 112
Argonne National Laboratory, 96
Armstrong, Arnold, 151n1
arthritis, crippled by, 59
*Atlantic Monthly*, 98
atomic bomb, 77
Australia, 152n6
Austria, women's status in science, 68

165

# About the Author

**Joyce Tang** teaches sociology at Queens College of the City University of New York <www.soc.qc.edu>. Her work focuses on social stratification and mobility.